Aerial Ascendancy
Volume II – The Rise of Flight in Austria-Hungary

Carl J. Bobrow

Publisher's Note
For our aviation books, please see our website at: **www.aeronautbooks.com**.
I am looking for photographs of the less well-known German aircraft of WWI to complete this series. For questions or to help with photographs you may contact me at **jherris@me.com**

Interested in WWI aviation? Join The League of WWI Aviation Historians (**www.overthefront.com**), Great War Aviation Society (**www.greatwaraviation.org**), and Das Propellerblatt (**www.propellerblatt.de**).

ISBN: 978-1-964637-18-1

Editor: Kimball Worcester
Design and layout: Jack Herris
Cover design: Gary Kroman
Digital photo editing: Jack Herris

Aeronaut Books
Books for Enthusiasts by Enthusiasts
www.aeronautbooks.com

This book is dedicated to Marcus Zelezny,

Historian, scholar, and guiding light in the exploration of aviation's early legacy.

Your knowledge, research, and remarkable contributions have enriched a field that few have approached with such passion and depth. When I sought a foothold in this challenging subject, your generosity and insight became an indispensable resource—freely shared and deeply appreciated.

May this dedication stand as a tribute to your enduring impact and serve as a beacon of recognition for your invaluable efforts.

With heartfelt gratitude and respect,

Carl J. Bobrow

Table of Contents

Acknowledgments

This work would not have been possible without the support, guidance, and contributions of the following individuals:

Andrei Alexandrov, John Anderson, Jon Barrett, Michel Bénichou, Lance Bronnenkant, Anders Bruun, Juraj Cervenka, Roger Connor, James Davilla, Klaas Dierks, George Fuderer, Jon Guttman, Von Hardesty, Jack Herris, Ted Huscher, Helmut Jagër, Anders Jonsson, Marat Khairulin, Carina Klemmer, Kees Kort, Gary Kroman, Victor Kulikov, David Layton, Koloman Mayerhofer, David Méchin, Stephen Miller, Chris Moore, Fred Murrin, Andrzej Olejko, Colin Owers, Robert Pandis, Gennadiy Petrov, Dan Polglaze, James Pratt, Harry Redner, Christian Reiterer, Sebastian Rosenboom, Gunnar Soderbaum, Olaf Stanger, Steve Suddaby, Hannes Täger, Pietsch Thorsten, Sarah Tisch, Jochen Vollert, Greg VanWyngarden, Gregory Vitarbo, Tobias Weber, John Weatherseed, Kimball Worcester, Marcus Zelezny.

Your encouragement, expertise, and generosity have been invaluable throughout this journey. I am deeply grateful for your inspiration and for being a part of this project.

Preface

Addressing early 20th-century events in Europe from a modern perspective presents unique challenges. Historical records and personal accounts of early aviation require a holistic approach. The historian's task involves sifting through disparate facts and details, weaving them into a narrative that resonates with modern readers while preserving the era's complexities. This process entails carefully reconstructing fragmented records, assessing them within their historical context, and assembling them into a coherent story that engages readers without sacrificing the era's intricacies. The result is an exploration that strives for historical relevance, guiding readers through a world whose contours, once vivid, have since faded and transformed.

Nations have faded and morphed, yet the lives and deeds of individuals retain their significance, holding a vital place in our understanding of the past. In delving into the achievements of early aviation pioneers, one finds an enduring relevance, as their efforts beneath the veils of a bygone era reveal deeper insights into the aspirations and technological ambitions that marked their time.

By the early 1900s, aviation—newly ascendant with the first powered flight in 1903—emerged alongside advances in dirigibles, whose development began decades earlier and was revolutionized by the adoption of gasoline engines in 1898. Though developed along distinct paths, both lighter-than-air and heavier-than-air technologies brought unique capabilities to the field, with dirigibles initially serving **in** reconnaissance and communication, and aeroplanes quickly demonstrating their speed and tactical agility. Together, they reshaped military doctrines and, in unforeseen ways, influenced geopolitical strategies. This series is not merely an academic exercise; it serves as a vital link to the foundations of today's technological and geopolitical landscapes.

As readers delve into the nuanced roles of aviation within the militaries of Russia, Austria-Hungary, and Germany—nations soon caught in a global conflict that would profoundly alter their societies—the question arises: what role did aviation play beyond popular imagination and nascent military planning? The promise of flight, initially viewed with awe and wonderment, quickly transformed into a military asset of significant consequence. Investigating how different designs, tactical deployments, and reconnaissance techniques were applied reveals aviation's transformation from a novel concept into a formidable military tool that fundamentally altered warfare's nature.

This series highlights often-overlooked aspects of early aviation, offering insights into its role in shaping the 20th century. By exploring the synergy between innovation in aviation and its gradual acceptance within military and societal frameworks, this narrative extends beyond historical recounting. It serves as a bridge, connecting past aspirations with present realities, inviting readers to reflect on the enduring legacy of this pioneering era.

Linguistic and Technical Practices

As we begin this exploration, it is essential to address some of the specific linguistic and technical practices employed throughout this book. The era in question was marked by diverse linguistic nuances, particularly in the military terminology of Russia, Austria-Hungary, and Germany. To preserve historical accuracy and enrich the narrative, Austro-Hungarian and German terms will first appear in their original language, followed by a translation in brackets. After the initial usage, we will primarily use the English translation to streamline the reading experience without compromising historical integrity. Occasionally, the original term will reappear, accompanied by its translation for clarity. Russian terms, which are pivotal to the narrative, are presented in transliterated format. To make them accessible to readers unfamiliar with Cyrillic script, we follow the BGN system, respecting both the historical authenticity and readability of the text. Names required careful handling, as relying on a family name alone could cause confusion when individuals shared surnames. Thus, full names are provided upon first mention, with family names used thereafter. Regarding military ranks and titles, we have included a chart with the ranks and military designations relevant to this work, along with their period-appropriate abbreviations where available or those widely accepted today. Additionally, the names of locations and military units have been drawn from contemporary records to ensure historical accuracy. Despite our rigorous research and verification efforts, it is possible that some variances or discrepancies in nomenclature may remain due to the vast scope of historical sources. We acknowledge any such unintentional errors, and we appreciate the reader's understanding. This approach to language and technical specificity is more than stylistic; it is essential for conveying the complexities and nuances of this historical narrative with precision.

Right: "First Flight Night at Steinfeldt," an aviation-themed carnival event at Wiener Neustadt, 12 February 1910.

Introduction

Austro-Hungary was an agrarian nation, dependent on grain production. Only 22% of the workforce was employed by industry. It was plagued by a divided political system with Austria and Hungary each having an equal role in the development of the military. These limitations meant that the country was in a poor position to develop and produce an air arm. Given its paucity of resources, it was remarkable that the country was able to accomplish as much as it eventually did.

Carl Bobrow's superlative work takes the reader through the slow and painful development of Austro-Hungary's nascent air arm. His book starts with the decision to produce a balloon force for reconnaissance. The Austro-Hungarian minister of war, Ferdinand von Bauer, created a delegation to investigate the use of LTA (lighter than air) aviation by military air arms around the world. He was assisted by balloonist *Oblt* Franz Hinterstoisser who performed some of the first ascents for the newly designated *k.u.k. Militar-Aeronautische Anstalt* [Military Aeronautical Institute]. The instability of the balloons was found to limit the ability to use binoculars or cameras. Later, the adoption of the stabilized Drachenballon improved this situation. In 1896, Hinterstoisser was given command of the *k.u.k. Militar-Aeronautische Anstalt*.

It was intended to use military balloons for artillery spotting. As a result, trainees were extensively trained in observation techniques. Due to the weight and lack of mobility the balloon forces had to remain attached to fixed positions in the fortress units. Subsequent advances in the equipment used to produce and store hydrogen gas would permit the balloon units to accompany field units by 1894.

By 1907 there were five *Festungsballonabteilung* (Nos. 2, 3, 4, 5, and 6) units assigned to the fortress balloon units. *Fesselballon-Abteilung* No. 1(a mobile unit) was created in 1898. This experimental unit performed successfully in military exercises later that year and led to balloon units being assigned to ground forces.

The narrative transitions from the balloon forces (which the military would soon conclude were too limited in the frontline reconnaissance role) to the army's newfound interest in powered airships. Unlike the balloons, powered airships could move along the front, permitting army commanders a more comprehensive view of the enemy positions. Here, however, the financial limitations of the dual monarchy would work to the advantage of the air service.

The military eventually concluded that airships were too expensive to pursue. After experimentation with a small series of airships further work was abandoned. Fortunately, the captive balloon program was allowed to continue for the time being. Thus, Austro-Hungary was spared the expense and delays in attempting to develop a force of dirigibles.

Even more important, the balloons would prove to be more useful than anticipated. Within two months of the start of the war, positional warfare with fixed lines would develop.

This situation benefited the balloon units which could provide critical information (at least at short range) in a timely fashion to frontline troops. Also, balloons would prove invaluable in regulating artillery fire against enemy positions. The contribution of these balloon units is often minimized in World War One histories. However, examining the daily French Armée aviation reports reveals that nearly 40% of the entries deal with the contributions of the balloon units.

The text describes the arduous efforts to establish a force of powered aircraft. It was hoped that these airplanes would permit a longer-range reconnaissance capability. To develop this new asset, in 1909 the *k.u.k. Militar-Aeronautische Anstalt* became the *Luftschifferabteilung* (Aeronautical Department). The base at Wiener Neustadt would serve as the main site for aeroplane development, while production would initially take place at Fischamend.

Aircraft were felt to be cheaper than powered airships and more effective than the balloon forces. The *Luftschifferabteilung* would have a chance to test this theory when a H.F.1*bis* was donated to them by a private citizen. Attempts to get the aircraft off the ground proved unsuccessful, and it was donated to a museum. In October 1909 a Wright aircraft was gifted to the military. This machine at least proved to be capable of flight. A second aircraft base was created at Rákos-mező. This was done to maintain the balance of the Hungarian side of the dual monarchy's aviation program.

With the potential for military contracts, there was now a resurgence of interest in aeronautics. Igo Etrich, Richard Knoller, and Ludwig Lohner would soon be providing effective aircraft to the military. For the time being, however, aircraft would be acquired through donations from private citizens.

In April 1911 the Austrian military purchased its first airplane - an Etrich Taube. This permitted the formation of the *Flugmaschinen Instruktionsabteilung* flying school.

Military exercises conducted between 27 August and 23 September 1911 gave the military operational experience with these new machines. The military pilots flew Etrich Taube airplanes, while the civilians used a combination of Taubes and their own personal aircraft. The results of these trials showed that the airmen performed effective reconnaissance flights.

The account offers a detailed picture of the other air meets, exercises, and aviation personalities of this time, all of which proved to be of critical importance in the development of the k.u.k. Air Service.

The 1912 Balkan War resulted in the decision to expand military aviation. 18 Etrich Taube and 32 Lohner *Pfeilflieger* were purchased by the end of 1912 .

As 1914 approached, the decision was made to expand army aviation to 15 Fliks which would require 90 aircraft and 120 pilots. However, with only the lone flying school at Wiener Neustadt in operation, there were not enough pilots to man these units - about 85 pilots. An attempt was made to train non-commissioned officers and enlisted men as pilots. With only 39 aircraft on strength, the only way to obtain further machines would be to order them from Germany.

As war began in 1914, the air service had to face the possibility of going to war against, France, Italy, Romanian, Russia, Serbia, and the United Kingdom with this meager force.

Assigned to patrol and perform bombing missions over the Adriatic and Mediterranean Seas, the k.u.k. naval air arm was in some respects more important than the army aviation service. From 1915 the flying boats of the navy, operating from Fiume, Pola, and Trieste, would be engaged in combat with the Italian aviation forces.

After abortive attempts to use captive balloons and operate aircraft from ships, the responsibility for these missions would fall on the navy's aeroplanes. From 1911 to June 1914 *Lschlt* Viktor Klobucar Rukavina von Bunic would serve as Kommandant of the k.u.k. Seeflugwesen. He made valuable contributions to the development of naval aviation prewar.

The study deftly covers the development of aircraft suitable for naval operations. One of the most important prewar acquisitions were the French Donnet-Lévéque flying boats. These were required because they were operationally proven and were superior to k.u.k. designs. Additional aircraft were procured prewar, and three copies were built at Pola in 1913. Their effectiveness led the naval command to conclude that flying boats were the best choice for naval missions. The flying boats were to be used for maritime reconnaissance, mine spotting and light bombing missions.

In 1914 there was a significant increase in funding for naval aviation; this led to acquisition of more types both from abroad, as well as those built at the Pola factories.

The narrative also highlights the contribution of such important aviation pioneers as Hptm Ferdinand Ritter Cavallar who commanded the Fischamend *Versuchsabteilung*, Richard Knoller, Richard von Mises, and many more. These gentlemen are forgotten today but are given their proper due in this work.

In conclusion, Carl Bobrow's *Aerial Ascendancy - The Rise of Flight in Austria-Hungary* gives the reader a chance to appreciate the work of the military and technical pioneers who created an air arm under extremely difficult economic political, and industrial conditions. While slow to appreciate the importance of aviation, the Austro-Hungarians avoided Germany's mistake of concentrating on dirigibles. The dual monarchy continued the development of its balloon forces, while adopting a far-thinking approach to the use of airplanes. While the country would never be able to overcome its inherent technical, economic, and political deficiencies, it was, in the end, able to develop effective army and naval air services.

James J. Davilla

Austro-Hungarian Army and Navy Officer & Enlisted Ranks

Army Rank	Abbreviation	USA Equivalent
Generalfeldmarschall	*GFM*	Field Marshal or General of the Army
Generaloberst	*GO*	Colonel General
General der Infanterie	*GdI*	General of Infantry
General der Kavallerie	*GdK*	General of Cavalry
Feldzeugmeister	*FZM*	General of Artillery
Generalleutnant	*Genlt*	Lieutenant General
Feldmarschalleutnant	*Fmlt*	Lieutenant Field Marshal
Generalmajor	*Genmjr*	Major General
Oberst	*Obst*	Colonel
Oberstleutnant	*Obstlt*	Lieutenant Colonel
Major	*Mjr*	Major
Hauptmann	*Hptm*	Captain
Rittmeister	*Rittm*	Cavalry Captain
Oberleutnant	*Oblt*	First Lieutenant
Leutnant	*Lt*	Lieutenant
Fähnrich	*Fhn*	Ensign
Offizierstellvertreter	*Offstv*	Acting officer
Kadett	*Kdtt*	Cadet
Stabsfeldwebel	*Stbsfw*	Staff Sergeant
Feldwebel	*Fw*	Master Sergeant
Unteroffizer	*Uffz*	Non-Commissioned Officer, Sergeant
Zugsführer	*Zgsf*	Master Corporal

Navy Rank	Abbreviation	USA Equivalent
Admiral	*Adm*	Admiral
Vizeadmiral	*VAdm*	Vice Admiral
Kontreadmiral	*KAdm*	Rear Admiral
Linienschiffskapitän	*Lkpt*	Captain
Fregattenkapitän	*Fkt*	Commander
Korvettenkapitän	*Kkpt*	Lieutenant Commander
Linienschiffsleutnant	*Lschlt*	Lieutenant
Fregattenleutnant	*Flglt*	Lieutenant Junior Grade
Fregattenfähnrich	*Ffhr*	Ensign
Korvettenfähnrich	*Kfhr*	Midshipman
Seefähnrich	*Sfr*	Naval Academy Midshipman
Kadett	*Kdt*	Naval Academy Candidate

Above: The Lohner B.I "*Pfeilflieger*," distinguished by its swept "Spanish Wings" [*Spanische Flügel*], was powered by a 90 PS Austro-Daimler engine and delivered under contract to the *k.u.k. Luftschifferabteilung* in early 1913. A total of 28 machines were procured, each bearing red-white-red identification markings on the fuselage and wing tips. Used primarily for officer instruction and aerial reconnaissance during maneuvers, the B.I demonstrated stable flight characteristics and speeds exceeding 130 km/h, laying the aerodynamic groundwork for subsequent Lohner developments.

1. Dawn of Aerial Endeavors in Austria-Hungary

Above: Viktor Silberer, captured in his younger years, exemplifies the pioneering spirit that drove early aeronautical advancements in Austria-Hungary. As an entrepreneur, journalist, and politician, Silberer was instrumental in advocating for the integration of aeronautics into the Austro-Hungarian military. *Credit: Marcus Zelezny*

Numerous scholars recounting the development of the *k.u.k.*[1] military aviation have emphasized that it was Silberer's exhibition that ultimately convinced the Austro-Hungarian political and military authorities to incorporate military aeronautics into the *k.u.k. Armée.*

By the late 19th century, intrepid civilian pioneers, military officers, and adventurers embarked on aerial journeys across Europe. The public observed and celebrated a series of sensational balloon ascents and expeditions. Viktor Silberer, a prominent Austro-Hungarian entrepreneur, journalist, writer, politician, and passionate aeronaut, was particularly noteworthy. Silberer embarked on his inaugural ascent in 1869 while in the United States. As a war correspondent during the Franco-Prussian War (1870–71), he witnessed the French deployment of balloons during the Siege of Paris, which solidified his conviction in the significance and potential of ballooning. In the summer of 1881, he ascended over Vienna with the esteemed French aeronaut Eugène Godard, inspiring him to purchase his own balloon from the Brissonnet company in Paris. With his newly acquired 1,100-cubic-meter balloon, named *Vindobona* (Latin for Vienna), Silberer made his first cautious ascent over the city on 10 August 1882. Over the next five years, he completed 187 untethered flights in this balloon, establishing his aeronautical expertise and the diverse applications of ballooning.[2]

In 1888, Silberer organized the *Ausstellung für Luftschifffahrt* [Aeronautics Exhibition] in Vienna, held from April to October 1888 at Silberer's Aeronautische [Silberer Aeronautics Institute]. His institute was the largest private facility of its kind worldwide, and the exhibition was attended by Emperor Franz Joseph I of Austria. The emperor's interest in aeronautics mirrored societal enthusiasm for the potential of flight. While Emperor Franz Joseph I had only a passing interest in aeronautics, it was a clever effort by Silberer to garner public attention by managing the emperor's visit, thus gaining significant media coverage. The military was intrigued by the possible applications of the balloon, particularly captive balloons, which were the primary focus of attention at the time. Their capacity to function as elevated observation posts for reconnaissance and artillery spotting, especially in fortress warfare, was driving their adoption across Europe. Silberer announced that he could locally manufacture balloons for the military to avoid dependencies on foreign countries. The deployment of balloons for mobile field warfare had yet to be fully realized, largely due to the cost and technical challenges of developing a transportable gas generation unit and the requisite ground support equipment.

The *k.u.k. Technisches und Administratives*

Right: Promotional Poster for Viktor Silberer's 1888 *"Große Ausstellung für Luftschifffahrt"* [Great Exhibition for Aeronautics] at his Aëronautische Anstalt [Aeronautical Institute] in the Prater, Vienna. This poster advertises the exhibition held from 1 April to the end of October 1888, which featured daily demonstrations and displays of various aeronautical advancements. Silberer's institute was instrumental in advancing aviation in Austria-Hungary, laying the groundwork for the nation's early aeronautical and military ballooning efforts. *Credit: Marcus Zelezny*

Militärkomitee [Technical and Administrative Military Committee] was tasked by Ferdinand von Bauer, the Austro-Hungarian war minister, to investigate military ballooning in use. A four-member delegation embarked on an international fact-finding mission to gather information on the state of the art of military aeronautics. The delegation visited the balloon workshops of Henri Lachambre and Gabriel Yon in Paris, the *Preußischen Luftschifferabteilung* [Prussian Balloon Detachment] in Germany, and

Above: Viktor Silberer's Ascension Site on the *Feuerwerkswiese* [Fireworks Meadow] in the Prater, Vienna, circa 1890. The balloon house and surrounding infrastructure were established by Viktor Silberer on the *Feuerwerkswiese*, an area leased from the *k.k. Obersthofmeisteramt* [Imperial and Royal Lord High Steward's Office, responsible for managing courtly and state properties] in 1882. This site supported both free and captive balloon ascensions, featuring two specialized platforms—one for free flights and another for tethered balloon experiments—alongside a workshop for producing various types of balloons. In 1890, the site gained historical significance as the birthplace of Austria-Hungary's military ballooning efforts, hosting the first *k.u.k. Militär-Aëronautischer Kurs* [Imperial and Royal Military Aeronautical Course]. *Credit: Rudolf Ster*

the Balloon School at Chatham in Great Britain. The resulting report advocated the establishment of a military aeronautical service. In accordance with this recommendation, volunteer officers and ground support personnel were trained and with Viktor Silberer's expertise enlisted to train the inaugural group at his *Aeronautische Anstalt* [Aeronautics Institute] in Vienna.

From April to August 1890, training was conducted using Silberer's two balloons, named *Vater Radetzky* and *Budapest*. Ascents were made primarily by *Leutnant (Lt)* [Lieutenant] Franz Eckert, *Lt* Franz Hinterstoisser, *Oberleutnant (Oblt)* [First Lieutenant] Hermann Hoernes, *Oblt* Hans Sojka, *Oblt* Josef Trieb, and *Oblt* Josef Watzek. Additionally, two non-commissioned officers (NCOs) and 24 enlisted men received training. Several successful ascents took place during the practical course, with a total of 28 ascents in 1890. A second course commenced in May 1891, with ascents primarily made by *Lt* Franz Hinterstoisser, *Oblt* Franz Eckert, *Oblt* Wilhelm Eyberger von Wertenegg, *Lt* Edmund Müller, *Oblt* Hans Sojka, and *Oblt* Josef Trieb. This course enrolled six officers, three NCOs, and 38 enlisted men. A minor incident occurred when a cross-country balloon flight landed in Russian territory, fortunately without political repercussions. The course was successfully completed, and subsequent annual courses proceeded as planned. Later training was performed by the military only, without Silberer's engagement. In the 1890 trainings, Silberer performed only three ascents because, shortly after, his severe rheumatoid arthritis prevented further direct engagements.

Endnotes

1. *K.u.k.*, kaiserlich und königlich, meaning "imperial (Austria) and royal (Hungary)," pertains to the Austro-Hungarian Empire 1867–1918.
2. Viktor Silberer, *Im Ballon! Eine Schilderung der Fahrten des Wiener Luftballons 'Vindobona' im Jahre 1882*, Verlag der Allgemeinen Sport-Zeitung, Wien, 1883 [In a balloon! A narrative of the flights of the balloon *Vindobona* in 1882].

2. The Establishment of k.u.k. Aeronautic Service

Above: Balloon ascent from the *Feuerwerkswiese* in the Prater, Vienna, on 5 June 1890. Among the participants was *Leutnant* Franz Hinterstoisser, whose presence is subtly underscored by his signature on this photograph. The ascent occurred in Vienna's renowned public park, the Prater, at a site later named "*Silberer-Platz*" to honor Viktor Silberer's contributions to aeronautics in Austria-Hungary. *Credit: Rudolf Ster*

Once a group of balloon aeronauts and ground support personnel had been trained, the *k.u.k. Kriegsministerium* [War Ministry] assumed responsibility for training before formally establishing a military aeronautical service. In August 1892, selected balloon aeronaut officers gathered in Vienna to finalize details for the *k.u.k. Militär-Aëronautische Anstalt* [Military Aeronautics Institute] under the guidance of *Oberst* (*Obst*) [Colonel] Philipp Hess and civilian Technical Counsel Dr. Friedrich Wächter, both from the *k.u.k. Technisches und Administratives Militärkomitee* [Technical and Administrative Military Committee].[1]

Simultaneously, an unused area in the Vienna Arsenal on the southern edge of the city was chosen as the site for the future training establishment, including the necessary ground infrastructure. Despite the barren landscape, the future military aeronauts had to make do with the location.[2]

By July 1893, the infrastructure work was completed, and the *k.u.k.* Military Aeronautics Institute was established with a balloon hangar, two material sheds, a gas generator plant, a small chemical laboratory, and a photographic space.[3] The inclusion of a photo-laboratory indicated an intent to utilize balloon photography for reconnaissance purposes.

However, experts debated the practical value of balloon photography due to limitations such as the distance and image quality, as well as the inability to develop photographs until the balloon had been roped down. Instead, visual reporting from the balloon basket was the preferred option. Various methods of communication were tested, with the wired telephone emerging as the optimal solution in 1905.[4]

The Prusso-German Army had already made significant advancements in balloon aeronautics before Austria-Hungary. After two balloons failed during the Franco-Prussian War (1870–71), the Berlin War Minister established *Ballon-Versuchsdetachement* [Balloon Test Detachment] for trial and testing on 9 May 1884. The unit was expanded and renamed the *Preußischen Luftschifferabteilung* [Prussian Balloon Detachment] on 8 May 1886.[5]

Like the Austro-Hungarians, the Germans recognized the tactical value of captive balloons in artillery fire observation and experimented with balloon photography. An exercise in 1892, involving daytime and nighttime photography, yielded lackluster results. However, photographic experiments continued despite the unfavorable outcomes.[6]

In July 1893, the newly established *k.u.k.* Military Aeronautics Institute procured its inaugural balloon from Germany, thereby initiating military service and training operations. The Austro-Hungarian Empire

Above: First military balloon hangar, Vienna Arsenal, 19 September 1895. This hangar represents the early stages of Austro-Hungarian military aeronautics. Established following recommendations from the *k.u.k. Technisches und Administratives Militärkomitee* [Technical and Administrative Military Committee], it served as the base for the newly formed *k.u.k. Militär-Aëronautische Anstalt* [Military Aeronautical Institute], where the first Austro-Hungarian balloon aeronauts were trained. *Credit: Marcus Zelezny*

chose to order German-produced balloon materials and ground equipment. Silberer was frustrated that the military did not order his balloons, despite his proven capability. He had manufactured three balloons in 1888, demonstrating his ability to produce them domestically.[7]

On 20 August 1893, the *k.u.k.* Military Aeronautics Institute was incorporated into the *1. Bataillon des Festungsartillerieregiments Kaiser Nr. 1* [1st Battalion of the *Kaiser* No. 1 Fortress Artillery Regiment], marking what would be the inception of the future *k.u.k. Luftschifferabteilung* [Imperial and Royal Aeronautics Department]. The organization's mission entailed monitoring and leveraging advancements in aeronautics while training officers and ground crew members, with annual military-aeronautical courses provided for this purpose.

Oblt Josef Trieb assumed the role of the first commander, assisted by *Oblt* Franz Hinterstoisser, one of the most renowned military balloon aeronauts in the *k.u.k. Armée* at the time. Hinterstoisser primarily accrued his expertise through military balloon ascents in Vienna from 1894 to 1900. Only in 1901 did he start performing civilian flights as a member of the *Wiener Aëro-Klub* [Vienna Aero-Club], which was later renamed the *Österreichischer Aero-Club* [Austrian Aero-Club]. Hinterstoisser was exceptionally talented, supported by Silberer, and had a strong relationship with Archduke Leopold Salvator, performing several ascents together. The initial staff complement of the *k.u.k.* Military Aeronautics Institute included two officers, four (NCOs), and 26 enlisted men for ground crew.[8] The inaugural training course commenced on 20 August 1893, catering to six officers, 60 NCOs, and enlisted men who volunteered from fortress artillery and technical troops.[9]

The *k.u.k.* Military Aeronautics Institute soon encountered significant aerodynamic issues with its *Kugelballons* [spherical balloons], a problem that was universally prevalent with all captive aerostats'

Above: Interior of a supply shed at the *k.u.k. Militär-Aëronautische Anstalt*, where a group of soldiers are meticulously inspecting and organizing a vast array of materials related to ballooning operations. The building serves as a vital point for the intake, processing, and storage of aeronautical supplies. *Credit: Rudolf Ster*

ascents. In gusty winds, captive balloons demonstrated considerable instability, with even low wind speeds of approximately 6 m/sec (13.5 mph) causing them to be forced to the ground—a critical limitation for military use.[10] Additionally, even under more favorable wind conditions, gusts caused excessive swaying of the balloon baskets, hindering observers' ability to obtain valuable reconnaissance data. Consequently, binoculars were seldom utilized, and cameras were practically unemployable.[11]

The issue of instability was a widespread challenge faced by captive balloon detachments across the globe. This prompted scientists, aeronautical pioneers and engineers to seek a solution. Beginning in 1888, German engineer-officers *Hauptmann (Hpt)* [Captain] Hans Bartsch von Sigsfeld, *Major (Mjr)* [Major] August von Parseval, together with August Riedinger began developing a unique balloon design to resolve the problem. The project was financially backed by Riedinger, the owner of *Ballonfabrik Augsburg* [Augsburg Balloon Factory]. In 1893 they produced a balloon that was capable of remaining stable even in strong, gusty winds.[12]

After the adoption of the *Drachenballon* by various European militaries, including the *k.u.k.* Military Aeronautics Institute, it became a widely accepted standard. As World War I progressed, the balloon's initial filling volume proved insufficient to reach higher altitudes for better observation as well as evading enemy anti-balloon artillery fire. Consequently, the *Drachenballon* was modified to accommodate 800 and 1,000 cubic meters of volume.[13]

Upon the establishment of the *k.u.k.* Military Aeronautics Institute , numerous military personnel underwent training, including officers, general staff officers, NCOs, and enlisted men. The primary focus of the training was artillery reconnaissance. Course participants developed expertise in observation and

Above: Balloon at the k.u.k. Military Aeronautical Institute, Vienna Arsenal, in the background. A large group of soldiers handles a military balloon, highlighting the challenges of early ballooning ground management. Whether preparing for ascent or bringing the balloon down, the image underscores the complexities and manpower required in Austro-Hungarian aeronautical operations during this period. *Credit: Rudolf Ster*

recognition of enemy positions during live balloon ascents. The 1894 course recorded approximately 200 ascents and 20 long-distance balloon journeys. This course also marked the first instance of testing artillery fire guidance from a balloon.[14] The specific means of air-to-ground communication employed remains unknown.

By 1896, Franz Hinterstoisser, promoted to *Hauptmann*, assumed command of the *k.u.k.* Military Aeronautics Institute from *Oblt* Trieb. Hinterstoisser maintained his position until 1912, save for a brief external assignment.[15] As a novel technology for the military, the aeronautical service initially lacked experience and employment principles. The development of knowledge and skills was a gradual process, culminating in the issuance of service regulations and employment principles for the training and utilization of military captive balloons in 1912.[16]

As part of the fortress artillery, captive balloons were confined to stationary deployment and served solely as an aid to fortress artillery. Major obstacles to mobility included the balloons' reliance on nearby gas-generating stations and the labor-intensive process of hauling balloons in by hand winch. Although horses and steam engines were tested for winching operations, only combustion engine-powered winches offered a feasible solution.[17]

In 1898, the *k.u.k. Armée* adopted the Parseval-Sigsfeld *Drachenballon*, two years after its introduction to the Prussian-German Army. That same year, the jubilee exhibition celebrating the 50th anniversary of Emperor Franz Joseph I's reign took place in Vienna. During this exhibition, a Parseval-Sigsfeld-Riedinger *Drachenballon* performed captive ascents, with Sigsfeld and Riedinger present. Following this demonstration, the military acquired the *Drachenballons*.[18] Concurrently, the personnel budget of the *k.u.k.* Military Aeronautics Institute was increased to encompass five officers and 62 NCOs/enlisted men.[19] Overcoming the primary hindrance to mobility, hydrogen gas production through gas generators and storage in steel cylinders became possible in 1894, enabling captive balloon detachments to accompany the field army.

Above: Balloon at the *k.u.k.* Military Aeronautical Institute, Vienna Arsenal, 20 August 1893. The Aeronautical Institute was officially incorporated into the 1st Battalion of the Kaiser No. 1 Fortress Artillery Regiment on this date, marking the foundation of what would become the *k.u.k. Luftschifferabteilung* [Imperial and Royal Aeronautical Department]. The institute's mission focused on advancing aeronautical knowledge and training personnel, with *Oblt* Josef Trieb and *Oblt* Franz Hinterstoisser playing key roles in its early operations. *Credit: Rudolf Ster*

Prior to establishing mobile balloon detachments, the Vienna War Ministry assigned fortress detachments, including equipment and ground personnel, to Krakau and Przemyśl from 1895 onward. Between 1897 and 1907, three additional fortress balloon stations were added to the existing detachments with updated equipment:

Festungsballonabteilung №. 2 in Krakau,
Festungsballonabteilung №. 3 in Przemyśl,
Festungsballonabteilung №. 4 in Pola,
Festungsballonabteilung №. 5 in Cattaro,
Festungsballonabteilung №. 6 in Trento.[20]

Upon the implementation of the *Drachenballon* in 1898, the War Ministry examined its mobile utilization by establishing the *Fesselballon-Abteilung* [Captive Balloon Detachment] №. 1 at Fischamend. Given that mobility necessitated manpower, the mobile detachment comprised a total of 63 individuals (seven officers, 12 NCOs, and 44 enlisted personnel), as well as 12 horse-drawn carriages. Motorization had yet to be introduced; thus, horses served as the primary means of mobility, with 56 horses in total, in addition to riding horses for officers.

The 12 carriages consisted of eight four-in-hand carriages, intended to expand to six-in-hand during wartime, and four two-in-hand carriages designated for the transportation of balloons and ground equipment. Six carriages were reserved for gas cylinders, while one carriage each was assigned for the balloon envelope and ropes. Crews rode on two carriages, with additional transport means provided for other materials and provisions. In the autumn of 1898, the detachment's

Above: Franz Hinterstoisser. A key figure in Austro-Hungarian military aviation, Franz Hinterstoisser played a significant role in the development of the *k.u.k. Militär-Aëronautische Anstalt* [Imperial and Royal Military Aeronautical Institute]. His leadership spanned crucial years in the early history of military aeronautics, culminating in the establishment of service regulations and principles for military balloon operations by 1912. *Credit: Rudolf Ster*

Above: This photograph captures nearly 20 uniformed personnel from the *k.u.k.* Military Aeronautical Institute engaged in a meticulous operation involving a *Kugelballon* [spherical balloon]. Utilizing a hand crank mechanism and coordinated teamwork, the scene highlights the disciplined practices of early military ballooning in Austria-Hungary. Whether preparing the balloon for ascent or managing its retrieval, the image, part of the "*Luftschifferkarte*" [Aeronauts' Card] series published by Diesel & Langer in Vienna, reflects the strategic importance of such aeronautical operations, with observers standing by to witness the procedure. *Credit: Rudolf Ster*

operational readiness was successfully assessed during an exercise.

Following the onset of motorization in the early 20th century, horse-drawn carriages were substituted with motor vehicles and the limber system, accompanied by motor winches. Throughout World War I, it became evident that the initial manpower and material allocation were insufficient. For comparison, a German *Feldballon-Abteilung* [Field Balloon Detachment] comprised approximately 250 individuals, including ten officers, nearly 200 horses, and 35 carriages, to deploy a single balloon.

Endnotes

1. Reinhard Karl Boromäus Desoye, *Die k.u.k.Luftfahrtruppe - Die Entstehung, der Aufbau und die Organisation der österreichisch-ungarischen Heeresluftwaffe 1912-1918* [The Austro-Hungarian Air Force—The emergence, structure and organization of the Austro-Hungarian Army Air Force 1912–1918], 15.
2. Desoye, 16.
3. Desoye, 17.
4. *Waffen-Arsenal Band-149, Deutsche Fessel und Sperrballone* 1900-1945 [Waffen-Arsenal Issue-149 "German Tethered and Barrage Balloons 1900-1945"], 10; Oliver Richter, *Feldluftschiffer: The German Balloon Corps and Aerial Reconnaissance Erlangen, Germany*: Verlag Jochen Vollert, Tankograd Publishing, 2013, 4–11; Klaus Held, 'Die Bayerische Ballon- und Luftschifferabteilung bis 1918' in *Blätter zur Geschichte der Deutschen Luft- und Raumfahrt XIX* [Pages on the history of aviation and astronautics XIX], Document ID: 82401004.

Above: A *Drachenballon* Model 98 in 1912 is being prepared for training purposes, with the *Luftschiff* [Airship] hangar at Fischamend visible on the left side. The designation 'Model 98' refers to the year of its adoption, marking it as an advanced kite balloon design known for stability in variable wind conditions. The operation of observation balloons, such as this one, was notably labor-intensive, requiring significant personnel to manage and deploy effectively. *Credit: Rudolf Ster*

5. Friedrich-Christian Stahl, *Die Militärluftfahrt bis zum Beginn des Weltkriegs 1914. Textband 1*, [Military aviation up to the beginning of the World War in 1914] The Mittler Berlin 1965 (Reprint from Orig. 1941), 8.
6. Helmut Jäger, *Erkundung mit der Kamera: die Entwicklung der Photographie zur Waffe und ihr Einsatz im 1. Weltkrieg* [Exploration with the camera: the development of photography as a weapon and its use in World War I], München, Venorion VKA, 2007, 35.
7. Ernst Peter, *Die k.u.k. Luftschiffer und Fliegertruppe Österreich-Ungarns 1794-1919* [The *k.u.k.* Austro-Hungarian airship and flying troops 1794–1919], Motorbuch Stuttgart 1981, 24; Desoye, 16.
8. Archduke Leopold Salvator (15 October 1863 in Alt-Bunzlau, Bohemia, to 4 September 1931 in Vienna), from the House of Habsburg-Lothringen (line Habsburg-Lothringen-Toskana), was an Austrian archduke and military leader. He joined the Imperial and Royal Infantry Regiment №. 77 in 1878, later serving in various artillery and infantry commands, and achieved the ranks of *Feldzeugmeister* and *Generaloberst*.
9. Peter, *Luftschiffer*, 24.
10. Waffen-Arsenal Band-149, 9.
11. Peter, *Luftschiffer*, 25; *Waffen-Arsenal Band-149*, 9; Richter, *Feldluftschiffer*, 4-11
12. The first German patent by August Riedinger, Hans von Sigsfeld №. 75731 was applied for on 10 October 1893 and granted 14 June 1894. The subsequent British patent GB-1895-18773 for the *Drachenballon* was assigned to August Riedinger on 17 October 1895; *Aeronautical Journal*, Vol. 1., №. 2, April 1897, 18; *Die Luftschiffahrt* [Airships], Hildebrandt, 1907, Hinterstoisser. *Der Drachenballon von Parseval und Sigsfeld* [The Parseval-Sigsfeld kite balloon]. Mitt. Geg. Art. Gen., XXIX Jahrg., 1898, Wien, 320–324.
13. Waffen-Arsenal Band-149, 16; Richter, *Feldluftschiffer*.
14. Peter, *Luftschiffer*, 25.
15. Peter, *Luftschiffer*, 25, 28, 31.
16. Peter, *Luftschiffer*, 31.
17. Georg Paul Neumann, *Die deutschen Luftstreitkräfte im Weltkriege* [The German air forces in the World War], Berlin: E. S. Mittler und Sohn, 1920, 15.
18. Peter, *Luftschiffer*, 25; Klaus Peters, *Zur Entwicklung der österreichisch-ungarischen Militärluftfahrt von den Anfängen bis zum Ende 1915*. [On the development of Austro-Hungarian military aviation from the beginning to the end of 1915] PhD. Diss. Wien 1971. (Typographic Manuscript) Österr. Staatsarchiv II 59.894, 22.
19. Peter, *Luftschiffer*, 26.
20. Desoye, 16.

Above: Mobile steam-powered balloon winch. Early attempts to improve the mobility of Austro-Hungarian captive balloons included the use of steam engines for winching operations. Despite these efforts, significant challenges remained, such as the reliance on nearby gas-generating stations and the labor-intensive nature of balloon handling. Ultimately, while steam-powered winches represented an innovation, only the advent of combustion engine-powered winches provided a truly feasible solution for the mobility and operational effectiveness of military balloons.

Above: Soldiers operating a mobile hydrogen gas generation system for inflating military balloons. The setup comprises two main units: the right unit produces hydrogen gas through a chemical reaction, while the left unit serves as a control and purification system, regulating the gas flow before transferring it to two large filler or 'nurse' bags. A hose drawing water from a nearby pond supplies both units, for cooling or purifying the hydrogen gas. The filler bags are critical for managing the controlled inflation of larger balloon envelopes, ensuring a safe and steady deployment of the highly flammable hydrogen gas. *Credit: Rudolf Ster*

Above: Postcard celebrating the *k.u.k.* Military Aeronautical Institute, Vienna Arsenal. This early 20th-century postcard commemorates the Austro-Hungarian military's aeronautical advancements. The left image shows a *Kugelballon* [spherical balloon] in front of the Vienna Arsenal, while the center features a *Drachenballon* [kite balloon] with the same iconic backdrop. On the right, a bust of Emperor Franz Joseph I underscores the imperial support for these pioneering efforts in military aviation. *Credit: Rudolf Ster*

Above: Parseval-Sigsfeld *Drachenballon* in the field with a mobile gas generator system. This image captures the logistical setup required for early military balloon operations, showcasing the mobile gas generator system used to produce hydrogen for the *Drachenballon*. The need for water in this process is evident from the pond, hoses, and pumps in the background. The scene illustrates the complex and resource-intensive nature of operating captive balloons in the field during this period. *Credit: Rudolf Ster*

Left: Interior of a storage facility at the *k.u.k.* Military Aeronautical Institute, illustrating the meticulous arrangement of balloon envelopes alongside wicker observation baskets. The presence of a large group of soldiers highlights the considerable manpower necessary to manage the significant size and weight of these envelopes, whether for organized storage, retrieval, or repositioning within the facility. *Credit: Rudolf Ster*

3. *Luftschifferabteilung*

Above: *Generalmajor* Leopold Schleyer, commander of the *Verkehrstruppenbrigade* [Transport Troops Brigade], played a pivotal role in the development of Austro-Hungarian military aviation. In 1909, under his leadership, the *k.u.k.* Military Aeronautical Institute in Fischamend was renamed the *Luftschifferabteilung* [Aeronautical Department], which expanded its responsibilities to include airship operations.

In 1909, the *k.u.k.* Military Aeronautics Institute in Fischamend was renamed the *Luftschifferabteilung* [Aeronautics Department] and integrated into the newly established *Verkehrstruppenbrigade* [Troop Transport Brigade] under *Generalmajor* (*Genmjr*) [Major General] Leopold Schleyer.[1] This commander championed the military aeronautical service, envisioning a significant role for it within the *k.u.k. Armée*. The newly formed Aeronautics Department retained its original mission with the added responsibility of managing airships. Consequently, additional personnel, material, and infrastructure were required, while fortress balloons remained under the jurisdiction of the artillery.

As the responsibility for airships grew, the Vienna Arsenal's southern edge became too constrained for the new detachment. Consequently, an airship hangar, workshops, and additional buildings for training, administration, and accommodations for support staff were necessary. This prompted a search for a more suitable location, ultimately leading to the selection of Fischamend, situated roughly 16 kilometers southeast of the former site. Consequently, the Vienna Arsenal's operations for airships were discontinued, and Fischamend became the new home for the airship components of the Aeronautics Department in 1909. However, balloon ascents still took place at the arsenal until 1914.

Simultaneously, in 1909, Wiener Neustadt was selected to host the aeroplane operations of the Aeronautics Department, leading to its dual-base presence in both Fischamend and Wiener Neustadt. Fischamend primarily focused on airships and aeroplane production and repair, while the airfield at Wiener Neustadt was primarily used for pilot training. Members of the *k.u.k. Luftschifferabteilung* sought to distinguish themselves from other *k.u.k. Armée* services and developed their own esprit de corps. Consequently, in 1910, Emperor Franz Joseph I sanctioned the use of a stylized balloon as a service insignia on the collar tabs (or *Kragenspiegel*) of the tunic. This emblem, as well as the eventual name *Luftfahrtruppen* [Aviation Troops], formally adopted in 1915, persisted until the conclusion of the war in 1918, including for the aviators who later joined the ranks.[2]

The rapid advancements in the airship sector, as well as in aeroplanes, necessitated a reorganization of the Aeronautics Department. Commencing on 1 October 1911, this restructuring was completed by spring 1912 and remained in effect until the onset of World War I.[3]

Concurrently, the War Ministry shifted its focus increasingly toward aeroplanes, recognizing that balloons had become obsolete and airships prohibitively expensive. Aeroplanes, along with being more affordable, offered numerous tactical and logistical advantages, which led most *k.u.k.* military decision-makers to favor them. Consequently, *Obstlt* Emil Uzelac was appointed as the new commander of the Aeronautics Department. Uzelac, an energetic individual with the necessary technological expertise, would ultimately become the driving force behind both this department and, later, the *k.u.k.* Aviation Troops.[4]

Above: Balloon hangar at the *k.u.k.* Military Aeronautical Institute, Fischamend. As Austro-Hungarian military aviation evolved, the need for expanded facilities led to the relocation of operations from the Vienna Arsenal to Fischamend in 1909. This hangar, part of the new infrastructure at Fischamend, was used for housing and maintaining military balloons, essential for training and operational readiness in the *Luftschifferabteilung*. Although operations shifted to Fischamend, balloon ascents continued at the Vienna Arsenal until 1914. *Credit: Rudolf Ster*

From 1912 onward, the *k.u.k.* Aviation Troops prioritized the use of aeroplanes. While the balloon force persisted, captive balloons were not expected to endure modernization. In 1913, the Vienna War Minister ordered the dissolution of the *k.u.k. Feldballon-Abteilung* [Field Balloon] units. However, due to the slower pace of change within the *k.u.k.* Monarchy, these units were still operational when the war began in August 1914, accompanying their artillery units into battle. This delay would prove fortuitous, as captive balloons later assumed unanticipated importance in directing artillery fire in the positional war.[5]

At the turn of the century, concepts for balloons equipped with propulsion systems were proposed, built, and tested. The primary obstacle was the engine, as the power-to-weight ratio remained inadequate for aerial use. Once this issue was resolved, engine-driven aeronautics garnered increasing attention. With streamlined hulls, these new vessels resembled ships rather than balloons, and the term "airship" was applied. The military advantages of airships seemed evident, foremost they could be controlled and steered in flight. Whether tethered or in free flight, balloons could only go up or down. Airships could traverse considerable distances and altitudes, providing crucial strategic reconnaissance. Moreover, they could transport substantial payloads, including bombs, photographic and wireless telegraphy equipment for long-distance communication, and reconnaissance while airborne.

Nevertheless, airships presented a significant drawback: they were sizable targets, vulnerable to enemy artillery. However, their lofty cruising altitudes rendered them seemingly impervious to attack. France, Germany, and Russia, followed closely by Austria-Hungary, anticipated that airship fleets would confer tactical advantages in times of war. As Germany pioneered the zeppelin concept, it was unsurprising that

Right: *Kragenspiegel* [collar tabs], introduced in 1910 for members of the *k.u.k. Luftschifferabteilung*, symbolized the affiliation of service members with Austro-Hungary's pioneering military aeronautical forces. Initially designated for personnel in balloon operations, these insignia became standard across all aviation units as the role of aeronautics expanded within the empire. Their design reflects the early focus on ballooning, serving as a visual marker of the wearer's connection to the foundational era of Austro-Hungarian military aviation. *Credit: Robert Pandis*

Below: Austrian aviator license № 50 issued to *k.u.k. Leutnant* Josef Flassig, dated 30 April 1912. *Fédération Aéronautique Internationale (FAI)* certified pilot under the auspices of the *Österreichischer Aëro-Club*. He received his brevet as a *Feldpilot* on 1 March 1912. Flassig is wearing the balloon *Kragenspiegel* introduced in 1910 for members of the *k.u.k. Luftschifferabteilung*. On 17 May 1913, Flassig survived a serious crash while serving as a passenger in a Lohner *Pfeil Doppeldecker* [Lohner Arrow Biplane] piloted by *Hauptmann* Deodatus Andrich. During a flight from Mostar towards Montenegro, the crew encountered severe turbulence, lost consciousness due to exhaustion, and crashed from a height of approximately 800 meters. Andrich suffered a fatal skull fracture, while Flassig was thrown clear of the wreckage, surviving with minor injuries and a concussion. Tragically, Flassig later perished on 16 August 1914 as a member of FliK 1 during the early stages of the war. He and his observer, *Oblt.* Engelbert Wolf-Engelbert, were killed when their Lohner type D biplane Lohner (named Don Quichotte), suffered a catastrophic wing failure. *Credit: Christian Reiterer*

FÉDÉRATION AÉRONAUTIQUE INTERNATIONALE.

ÖSTERREICH.
AUTRICHE. Nr. 50

Die unterzeichnete von der **Fédération Aéronautique Internationale für Österreich** anerkannte Sportmacht bescheinigt, daß

Nous soussignés, pouvoir sportif reconnu par la **Fédération Aéronautique Internationale pour l'Autriche,** certifions que

Herr / Monsieur k. u. k. Leutnant Josef Flassig

geb. in / né à Wien am / le 13/5. 1886

nach Erfüllung aller von der **F. A. I.** verlangten Bedingungen diplomiert wurde als

ayant rempli toutes les conditions imposées par la **F. A. I.** a été breveté

Aviatischer Pilot — Aviateur

WIEN, am / le 30. April 1912

ÖSTERREICHISCHER AËRO-CLUB

Präsident: / Le Président:

Schriftführer: / Le Secrétaire:

Unterschrift des Inhabers:

Above: *k.u.k.* Military Aeronautical Institute at Fischamend. A *Kugelballon* is being prepared in front of the newly constructed, massive dirigible hangar, one of the largest structures at Fischamend. This hangar, built to house Austro-Hungarian airships, underscores the strategic importance of Fischamend as the central hub for military aviation development. In the background, a Lohner-Daimler *Pfeilflieger* [Arrow Flyer] further illustrates the diversity of aeronautical activities taking place at this critical site. *Credit: Rudolf Ster*

many in Vienna advocated for an airship fleet like those developed by the Germans, French, and potentially the Russians.

The decision-makers in the Vienna War Ministry expressed reservations about airship fleets, primarily due to the immense costs associated with their procurement, operation, and ground infrastructure. *Chef des Generalstabs* [Chief of the General Staff] Franz Conrad von Hötzendorf championed the establishment of a *k.u.k.* airship fleet, but his efforts were met with resistance in the War Ministry.[6]

It was only in the fall of 1909 that the Imperial War Ministry's negative stance toward airships shifted to cautious approval. On 7 October 1909, Emperor Franz Joseph I decreed, through his Very Highest Order, that the responsibilities of the Military Aeronautics Institute would be expanded to include airships. This marked the beginning of the brief and ill-fated chapter in the history of *k.u.k.* airships.

A prevailing narrative asserts that the emperor's admiration for airships directly influenced *General der Infanterie (GdI)* [General of Infantry] Baron von Schönaich's shift in attitude toward incorporating such vessels into the *k.u.k. Armée* . This perspective is partly founded on the emperor's attendance at an airship demonstration in Vienna on 16 October 1909, where he expressed his appreciation for the Renner brothers' personally funded dirigible *Estaric.*[7] However, while Emperor Franz Joseph I did have influence, in this instance, his opinion on the subject was not pronounced, and he did not exert his influence either way. The key figures in the decision-making process were Baron von Schönaich and *Genmjr* Schleyer of the Troop Transport Brigade. Evidence shows that the integration of an airship service into the *k.u.k. Armée* was already under consideration months prior to the Renner brothers' event, influenced significantly by Germany's aviation policies and practices.[8]

Above: The *Estaric*, a dirigible designed and constructed by the Renner brothers in Vienna, 1911, exemplified the ingenuity and ambition of Austro-Hungarian private aeronautical pioneers. This airship garnered significant attention, highlighting the potential of privately developed airships to contribute to the evolving landscape of aeronautics. Despite its private origins, the *Estaric* underscored the broader possibilities for airship technology within Austro-Hungarian aviation during this formative era.

Indeed, Franz Joseph had ordered the reorganization of the Military Aeronautics Institute into the Aeronautics Department with expanded airship responsibilities on 7 October 1909, with completion set for 17 October. Such significant decisions are seldom spontaneous, implying prior deliberation and resolution. Moreover, the assignment of this unit to the Troop Transport Brigade under *Genmjr* Leopold Schleyer, a known advocate for air service, further suggests a shift in the War Minister's official stance.

The restructure marked the military aeronauts' transition from experimentation to a recognized service. Consequently, the Aeronautics Department expanded its personnel to comprise four officers, 178 NCOs, and enlisted men. To assess European powers' military preparedness in airship development, a delegation was dispatched, with the central question concerning Austria-Hungary's method of procuring its first airship.

Two *k.u.k. Oberleutenant*s, Franz Adolf von Berlepsch and Franz Mannsbarth, traveled to Frankfurt am Main, Germany, for "airship commander training."[9] The Vienna War Ministry's imminent purchase of its first a non-rigid dirigible, a German Parseval-type, was clear. A facility was constructed in Fischamend to accommodate the airship.

Austria-Hungary lacked an industrial foundation for airship production, but the Vienna War Ministry sought self-sufficiency. Consequently, it enlisted the prominent Austrian industrialist Camillo Castiglioni to obtain licenses and build a *Pralluftschiff* [Parseval airship].[10] Castiglioni's passion for aviation, together with his status as a licensed balloon aeronaut and access to industry, made him the appropriate partner for the project.[11] To fund the airship's construction, Castiglioni established the Motor-Luftfahrzeug-GmbH on 23 April 1909, a joint subsidiary of the Österreichisch-Amerikanische Gummiwerke AG, Wimpassing, [Austrian-American Rubber Factory],

Above: At the *k.u.k.* Military Aeronautical Institute in Fischamend, the Parseval *M.I* airship is positioned outside the large dirigible hangar, reflecting Austria-Hungary's early advancements in military aviation. The Parseval *M.I*, Austria-Hungary's first military airship, was a product of collaboration between Austrian industrialist Camillo Castiglioni's Motor-Luftfahrzeug-GmbH and key Austrian industrial partners. This airship marked a significant step towards achieving self-sufficiency in airship technology for the Austro-Hungarian military, highlighting Fischamend's strategic role in the development and testing of military aeronautical innovations. *Credit: Rudolf Ster*

and the Österreichische Daimler-Motoren AG, Wiener Neustadt [Austro-Daimler Engines]. The Österreichische Daimler-Motoren AG supplied and built the gondola and engine, while the Österreichisch-Amerikanische Gummiwerke AG supplied and built the envelope. The first Austro-Hungarian military Parseval-type airship was assembled in Fischamend, deemed ready for acceptance by November 1909, and designated *M.I.* Following its successful maiden flight on 26 November 1909, *M.I* served various training and testing purposes, with numerous documented voyages.[12]

In 1910, the War Ministry commissioned a second airship, opting for a licensed build of a French semi-rigid airship designed by the Lebaudy Frères company, with Henri Albert Julliot as its chief engineer. Lebaudy-type airships had served as France's initial military airships since July 1905. The Austro-Hungarian Lebaudy ship was a product of Motor-Luftfahrzeug-GmbH, manufactured by the joint effort of the Österreichisch-Amerikanische Gummiwerke AG, which supplied the envelope, and the Österreichische Daimler-Motoren AG, which supplied the gondola and engine. Fischamend was the construction and assembly site for the Lebaudy airship, which underwent acceptance tests in 1910.[13]

The test flights were unsatisfactory, and only after a year of technical modifications did the airship enter service as *M.II* in February 1911. It remained in testing status until its decommissioning in 1912, after which it was used for airship personnel training.[14] In addition, *M.I* was decommissioned in 1912 due to high maintenance costs and the extensive ground support required.[15]

Despite the expenses, the *k.u.k. Armée* acquired a third airship, *M.III*, in 1911. This airship, of Austrian origin, was a non-rigid type like the *M.I*, designed by engineers Alexander Cassione, Wilhelm Stratmann, and a German engineer Nikolaus Basenach. The Österreichische Maschinenbau AG, Körting, and Österreichisch-Amerikanische Gummiwerke AG constructed *M.III*, which necessitated a second hangar in Fischamend due to its larger size.[16]

M.III, also known as the *Körting Luftschiff*, was powered by two 75-hp Körting engines and

Right: Camillo Castiglioni, an Austrian financier and one of Europe's wealthiest men in the early 20th century, played a key role in the development of aviation in Austria-Hungary. Born in Trieste in 1879, he rapidly advanced in the business world, becoming director of Österreichisch-Amerikanische Gummiwerke and later chairman of Austro-Daimler. Castiglioni co-founded the Wiener Aero-Club and established the Motor-Luftfahrzeug-Gesellschaft (MLG) in 1909, acquiring major aeroplane manufacturers like Brandenburgische Flugzeugwerke and Phönix Flugzeugwerke. His contributions significantly shaped the aviation industry during World War I.

Below: *Oberleutnant* Hans Hauswirth, airship pilot with the *k.u.k. Luftschifferabteilung*, poses with crew members beside the gondola of a Parseval dirigible. As one of Austro-Hungary's early airship pioneers, Hauswirth played a key role in developing operational practices that expanded the role of dirigibles within the empire's aeronautical efforts.

fulfilled the *k.u.k.* War Ministry's requirements for rapid transportability in a disassembled state. With aerodynamic improvements, *M.III* boasted enhanced range, a cruising speed of 60 km/h, superior maneuverability, and a greater payload capacity compared to *M.I* and *M.II*. Furthermore, it featured a wireless unit built by the Telefunken company.[17]

However, since 1910, a fourth airship, the *Stagl-*

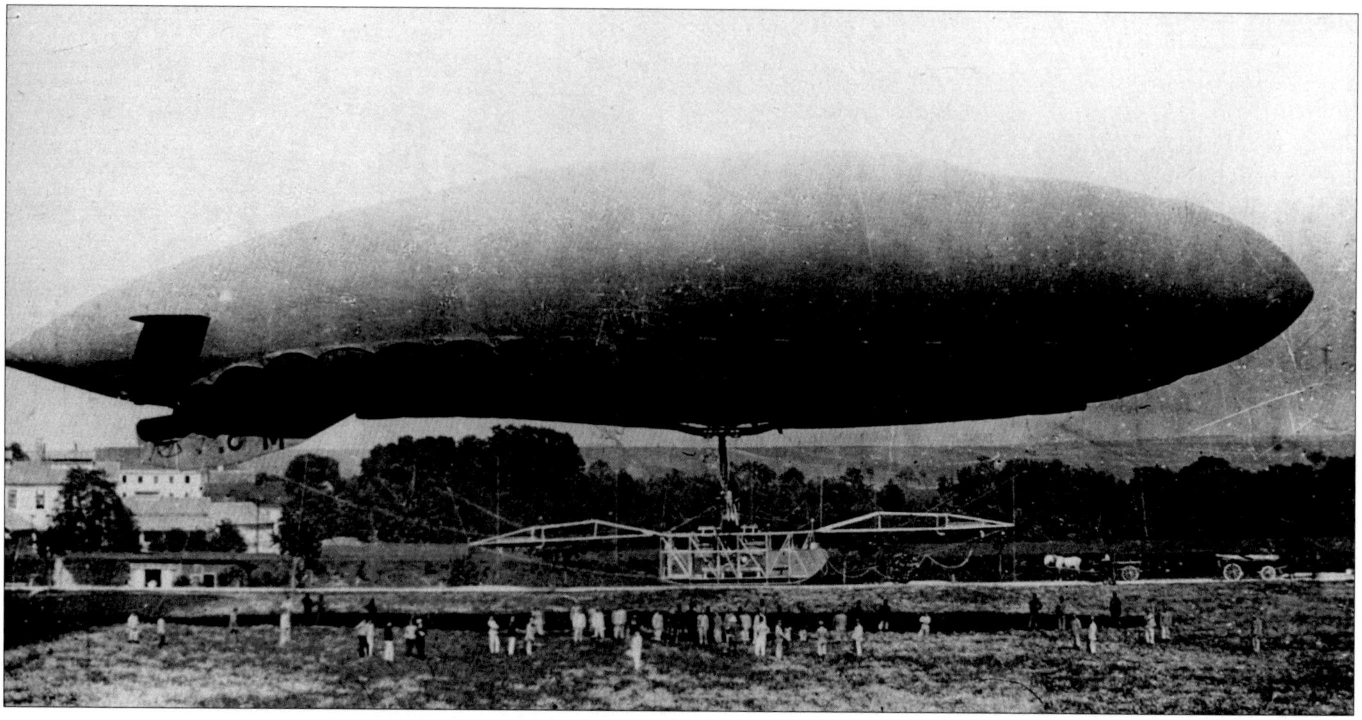

Above: The *Körting Luftschiff*, designated *M.III*, represented Austria-Hungary's third military airship. Designed by engineers Alexander Cassione, Wilhelm Stratmann, and Nikolaus Basenach, this non-rigid airship was powered by two 75 hp Körting engines and featured advanced aerodynamic improvements, a cruising speed of 60 km/h, and a wireless unit by Telefunken. Constructed in 1911, M.III was notable for its enhanced range, maneuverability, and payload capacity, reflecting the Austro-Hungarian military's commitment to advancing airship technology.

Mannsbarth-Lenkballon [Stagl-Mannsbarth Dirigible], had been present in Fischamend but was never acquired by the army. *Ingenieur* [Engineer] Hans Otto Stagl and airship officer *Hptm* Franz Mannsbarth designed the ship, which was privately funded. It was manufactured by Österreichisch-Amerikanische Gummiwerke AG, which supplied the envelope, and the Österreichische Daimler-Motoren AG, which supplied the gondola and engine. At 91 meters in length and a volume of 8,100 cubic meters, it was the world's largest non-rigid airship at the time. Powered by two 150-hp Daimler engines, achieving a speed of 68 km/h, the airship accommodated up to 30 people, with a service ceiling of 2,000 meters and eight hours of fuel.[18]

The Mannsbarth investment failed due to the War Ministry's expenditures on the three previous airships and a shift in favor of aeroplanes, which led to the phasing out of airships. As a result, the investors in the Stagl-Mannsbarth Dirigible were unable to recover their investment, and the airship was disassembled in 1914.[19]

In a peculiar turn of events, the *k.u.k.* Aeronautics Department acquired a fourth airship, *M.IV*, in 1912. The war minister accepted the airship as a donation from a wealthy *k.u.k* officer, *Hptm* Friedrich Boemches of the *Pionierbataillon 5* [Pioneer Battalion 5]. The *M.IV* was of Austrian origin, constructed by several national companies, measuring 57.5 meters in length with a 2,700 cubic meter volume. Powered by two 35-hp Körting engines, it had a cruising speed of 46 km/h, carried a crew of two to six, and had fuel for ten hours of operation. However, *M.IV* was decommissioned shortly after its first test flight.[20]

Only the Körting *M.III* remained in service at Fischamend, used for testing and trialing airborne equipment. The airship met a tragic end on 20 June 1914, eight days before the assassination of Archduke Franz Ferdinand and his wife, Sophie, Duchess of Hohenberg, in Sarajevo. On that day, *M.III*, under the command *Hptm* Johann Hauswirth took off from Fischamend to test the Scheimpflug-Kammerer aerial photographic apparatus.[21] The crew included *Oblt* Ernst Hofstätter, *Lt* Otto Haidinger, *Korporal* Franz Chadima, and *Gefreiter* Franz Weber. Additionally onboard were *Oblt* Adolf Breuer of the *k.u.k. Militärgeographisches Institut* [Military Geographical Institute] and *Ingenieur* Gustav Kammerer, of the Scheimpflug-Kammerer Institute for aerial photographic equipment, who were responsible for the photographic tests.

Above: The *Stagl-Mannsbarth Lenkballon*, photographed here in 1911, was the largest non-rigid airship of its time, measuring 91 meters in length with a volume of 8,100 cubic meters. Designed by *Ingenieur* Hans Otto Stagl and *Hptm* Franz Mannsbarth, this privately funded airship was powered by two 150 hp Daimler engines, allowing it to reach speeds of up to 68 km/h. Despite its impressive capabilities, the airship was never acquired by the Austro-Hungarian Army and was ultimately disassembled in 1914 due to shifting military priorities toward aeroplanes.

During the flight, strong winds caused *M.III* to drift off course. *Oblt* Ernst Flatz and *k.u.k Leutnant zur See* [navy lieutenant] Wolfgang Puchta took off in a Farman HF.20 biplane, presumably to investigate *M.III*'s drift, though their reasons remain unclear. Flatz circled around *M.III* multiple times, and on the third circle, the Farman biplane's wings collided with the airship, damaging its envelope. The aeroplane spiraled out of control, and *M.III* erupted in flames, crashing from a height of 400 meters and killing all crew members. Flatz and Puchta also perished in the aeroplane crash.[22] This catastrophic event marked the end of the brief era of *k.u.k.* military airships.

Endnotes

1. Leopold *Frh* [Baron] Schleyer von Pontemalghera (1858–1920), was born in Königgrätz, Bohemia, and died in Vienna. He graduated from the *k.u.k.* Technical Military Academy in Vienna and advanced through military ranks: *Lt* (1879), *Oblt* (1884), *Hptm* (1889), *Mjr* (1895), *Obstl* (1897), *Obst* (1900), and *Genmjr* (1907). He was promoted to *Feldmarschalleutnant* [Lieutenant Field Marshal] in 1912 and took command of the Troop Transport Brigade, overseeing developments in military transport and aviation. In 1915, he was further promoted to *Feldzeugmeister* [General of Artillery]. Schleyer was a recognized expert in transportation and communications and made significant contributions to military aviation, introducing wireless telegraphy in the army. He organized the first official Austrian airfield in Wiener Neustadt in 1909 and served as president of the Austrian Aeronautics Society. Ennobled in 1912 and later elevated to the title of von Pontemalghera, he was honored with titles, including Secret Counsel, and received an honorary doctorate from the Technical University of Vienna.
2. Robert S. Pandis, *Flight Badges of the Central Powers, 1914–1918, Volume II - The Imperial German Naval, Austro-Hungarian, Bulgarian, and Ottoman Turkish Air Services*, Imperial House Antiques LLC.,132: Jörg C. Steiner, 'Das Feldpiloten-Abzeichen' [The field pilot's badge], *Militärhistorische Themenreihe* [Thematic series], Band 4, 1–8.
3. Desoye, 32.
4. Peter, *Luftschiffer,* 298; The *k.u.k.* Aeronautics Department was reorganized on 18 June 1915 to become the *k.u.k.*

Above: The *Scheimpflug-Kammerer* aerial photographic apparatus, was undergoing testing aboard the *Körting Luftschiff M.III* in June 1914. The camera system, capable of simultaneously capturing seven photographs arranged in a radial pattern to create a panoramic view, was developed for aerial reconnaissance capabilities by providing detailed photographs for mapping and intelligence purposes. Unfortunately, the mission ended in tragedy when the *M.III* collided with a Farman HF.20 biplane near Fischamend.

Luftfahrtruppen [Imperial and Royal Aviation Troops], serving as the aviation arm of the Austro-Hungarian Empire's military. Both the *k.u.k.* Aeronautics Department and the subsequent *k.u.k.* Aviation Troops played significant roles in reconnaissance, aerial photography, pursuit, and bombing missions during World War I; Desoye, 98–99.

5. Desoye, 18.
6. Desoye, 18.
7. Anton Leopold Schuller, *Renner, Franz (1866–1912), Artist und Flugpionier. In: Österreichisches Biographisches Lexikon 1815–1950* (ÖBL). Band 9, Verlag der Österreichischen Akademie der Wissenschaften, Wien 1988, ISBN 3-7001-1483-4, S. 79–80; Brothers Alexander, Anatol Renner, *Wiener Luftschiffer-Zeitung* [Vienna aeronaut newspaper] ⊠. 1, 15 Marz 1914, XIII Jahrgang, 19; During a tour of the USA in 1907, the Renner family observed the Baldwin airship, which influenced their design of the *Estaric I*. This exposure helped shape the Renner brothers' airship design and construction. At the 1907 St. Louis Air Meet, Thomas Scott Baldwin demonstrated his dirigible, the *California Arrow*, powered by a Curtiss V-8 motorcycle engine.
8. Desoye, 19.
9. Reinhard Keimel, *Österreichs Luftfahrzeuge: Geschichte der Luftfahrt von den Anfängen bis Ende 1918*, [Austria's

Facing Page: A detailed report on the *Körting Luftschiff M.III* crash, published in the newspaper *Das interessante Blatt* on 25 June 1914. The article recounts the tragic collision between the *M.III* airship and a Farman HF.20 biplane near Fischamend during the airship's flight to further evaluate the *Scheimpflug-Kammerer* system. The accident resulted in the complete destruction of the airship and the loss of all crew members, including several prominent figures in Austro-Hungarian military aviation. This catastrophe marked a turning point, prompting the military to abandon its airship program and shift its focus entirely to aeroplanes.

Abonnementspreise mit wöchentlicher Postversendung: für Oesterreich-Ungarn: vierteljährig 3 Kronen 20 Heller, halbjährig 6 Kronen 40 Heller, ganzjährig 12 Kronen 80 Heller; für Deutschland: vierteljährig 4 M., halbjährig 8 M ganzjährig 16 M.; für alle übrigen Länder des Weltpostvereines: vierteljährig 7 Francs, halbjährig 14 Francs, ganzjährig 28 Francs. — Einzelne Nummern 24 Heller.

Redaktion und Administration: Wien, III. Rüdengasse 11 (Telephone 4199, 9767 und 9540), Stadtbureau: Wien, I. Schulerstraße 18.

Nr. 26. Erscheint Jeden Donnerstag. Wien, 25. Juni 1914. Abonnements durch jede Buchhandlung und Postanstalt. XXXIII. Jahrg.

Große Fliegerkatastrophe bei Fischamend.

Zusammenstoß eines Lenkballons mit einem Aeroplan. — Neun Todesopfer.

1. Die Trümmer des zerstörten „Körting"-Lenkballons werden von der Unfallstelle weggeschafft. 2. Die Ueberreste des verbrannten „Farman"-Doppeldeckers. 3. Der Führer des Körtingballons Hauptmann Hans Hauswirth. 4. Oberleutnant Ernst Hofstätter. 5. Vormeister Franz Weber. 6. Korporal Franz Chadima. 7. Zivilingenieur Gustav Kammerer. 8. Der Führer des Aeroplans Oberleutnant Ernst Flatz. (Siehe Seite 12.)

Nach photographischen Aufnahmen.

Der Kaiser bei der Schönbrunner Fahnenweihe. — Der Zar in Rumänien. — Gefechtsbilder aus Durazzo.

aeroplane: the history of aviation from the beginning to the end of 1918] H. Weishaupt Verlag, 1981, 50–65: Peter, *Luftschiffer,* 36.

10. Marcus Zelezny, *Aviation in Wiener Neustadt (1909–1912)*, Stanger Verlag, 2022, 195.
11. Camillo Castiglioni was an Austrian financier, banker, and philanthropist, one of the wealthiest men in Europe in the early 20th century. He was married to Anna Körting, the daughter of German engineer and industrialist Ernst Körting. He was born in Trieste in 1879. His aspiration to advance his societal position combined with his aptitude for foreign languages and financial matters helped drive his business acumen for international finance and the management of capital. Castiglioni was sent to Constantinople to represent the Österreichisch-Amerikanische Gummiwerke company in the production of tires. His successful efforts garnered him a position as head of the export department back in Vienna in 1902. He quickly advanced and became deputy director and then as director in 1909 for the large company. Additionally, he went on to become the chairman of the board for Austro-Daimler in Wiener Neustadt. In 1901, together with Viktor Silberer and Franz Hinterstoisser, Castiglioni was a co-founder of the Wiener Aero-Club, later the Österreichischen Aero-Club. In 1909, he bought his own balloon and successfully passed the aeronaut's exam on 24 August. Castiglioni founded the Motor-Luftfahrzeug-Gesellschaft mbH in 1909. The company acquired the Etrich patents and the distribution of the Etrich *Taube* in Austria-Hungary. In 1912, he was a partner in the founding of Ungarische Flugzeugwerke AG (Ufag) along with other aeroplane companies. Castiglioni acquired the aeroplane companies Brandenburgische Flugzeugwerke GmbH and the Hansa-Flugzeug-Werke Hamburg. In 1914 he merged the two with his own company, Deutsche Aero-Gesellschaft AG, renamed *Hansa- und Brandenburgische Flugzeugwerke AG*, with Ernst Heinkel as its chief designer. In 1917, Phönix Flugzeugwerke was also added to the group of aeroplane companies he owned.
12. Keimel, *Österreichs* 50–65: Peter, *Luftschiffer,* 36.; Zelezny, *Neustadt*, 195.
13. Keimel, *Österreichs* 50–65: Peter, *Luftschiffer,* 37; Neumann, *Internationalen Luftschiffe,* 94.
14. Keimel, *Österreichs* 50–65: Peter, *Luftschiffer,* 39.
15. Desoye, 21; Peter, *Luftschiffer,* 37.
16. Keimel, *Österreichs,* 50–65.
17. Keimel, *Österreichs,* 50–65; Peter, *Luftschiffer,* 41.
18. Keimel, *Österreichs* 50-65: Peter, *Luftschiffer,* 43; Desoye, 13; *Wiener Luftschiffer-Zeitung* [*Vienna Aeronaut* newspaper] ⊠. 8, 15 Juni 1914, XIII Jahrgang, 139–140.
19. Keimel, *Österreichs* 50–65; Peter, *Luftschiffer,* 46; Zelezny, *Neustadt*, 202.
20. Keimel, *Österreichs* 50–65; Peter, *Luftschiffer,* 46.
21. Theodor Scheimpflug, *Kapitänleutnant* of the *k.u.k. Kriegsmarine*: After concluding his service, Theodor Scheimpflug established an institute for aerial photography, which was managed after his death in 1911 by his collaborator, Engineer Gustav Kammerer. Kammerer, who was affiliated with the experimental department of the aviation division in Fischamend, conducted extensive testing of the aerial camera developed by the institute. Both Kammerer and the Scheimpflug-Kammerer aerial photographic device were tragically lost in the crash of the *M.III* airship on 20 June 1914, as documented in *Das Lichtbildwesen der öst.-ungarischen Luftfahrtruppen* 1914/18, 8. Consequently, no Scheimpflug-Kammerer aerial photographic apparatus was ever delivered to the Aeronautics Department for operational use.
22. Viktor Silberer, *Luftschiffahrt. Der Massenmord in Fischamend*, [Airship: mass murder in Fischamend] *Allgemeine Sport-Zeitung* [General sport newspaper], 28 June 1914, 679–680; Peter, *Luftschiffer,* 99; Stefan Stachniß, *Vorarlberger Soldaten bei den k. u. k. Luftfahrtruppen Österreich-Ungarns im Ersten Weltkrieg* [Vorarlberg soldiers with the Austro-Hungarian Aviation Corps in the First World War] (Masters Thesis)] Innsbruck, 2021, 26–29.

4. First Aeroplanes

Above: Hart Ostheimer Berg and Wilbur Wright, pictured in 1908, were pivotal figures in the promotion of the Wright Flyer in Europe. Berg, representing Charles Flint & Company, visited the Vienna War Ministry in September 1908 to propose the sale of the Wright Flyer to the Austro-Hungarian military. Despite the growing international recognition of the Wright brothers' achievements, including their successful flights in France that catapulted aviation into the modern age, the Austro-Hungarian War Ministry hesitated in purchasing the machine. The Wright Flyer had already begun to revolutionize aviation, demonstrating unprecedented control and stability, which influenced aviation across Europe.

Above: The shop sign of Österreichische Daimler-Motoren-Gesellschaft [Austro-Daimler Motor Company] at Wiener Neustadt. Established as a key player in the development of the Wiener Neustadt airfield in 1909, Austro-Daimler contributed significantly to both military and civilian aviation in Austria-Hungary. The company, under the leadership of figures like Ferdinand Porsche, played a crucial role in providing the technical expertise and resources necessary for the establishment and growth of the airfield, which became a central hub for aviation activities before World War I.

Around a decade later than other major European military powers, the Vienna War Ministry established a *k.u.k.* military balloon aeronautics unit. Rapidly catching up, they shifted focus to airships and aeroplanes, culminating in the establishment of a now aeroplane-centric *Luftschifferabteilung* in 1912. Prior to this, the *Königlich-Preußische Fliegertruppe* [Royal Prussian Aviation Troops] had already been established in Germany on 1 October 1912. The German Army had begun training military pilots and established their first military airfield at Döberitz as early as May 1910, setting a precedent that would later be followed by Austria-Hungary.[1]

In September 1908, Hart Ostheimer Berg of the Charles Flint & Company, who were representing the Wright brothers' company, visited the Vienna War Ministry to promote the sale of a Wright Flyer. However, the war minister, displaying typical Austro-Hungarian procrastination, decided not to purchase any aeroplanes until they were tested and deemed fit for war.[2]

In June 1909, the establishment of a military and civilian airfield in Wiener Neustadt was propelled by a coalition of influential figures and organizations. Key military advocates included *Genmjr* Leopold Schleyer, head of the, and *Hptm* Franz Hinterstoisser, commander of the *k.u.k.* Aeronautics Department. Their military support was crucial in gaining approval for the project.

Above: Hangars at Wiener Neustadt airfield. By 1912, Wiener Neustadt had become a dual-purpose airfield serving both military and civilian aviation. Initially established with the support of local municipal leadership and various military, civilian, and commercial interests, the airfield rapidly developed into a significant aviation hub. The community constructed and rented out these hangars to accommodate the growing aviation industry. By 1913, the military had fully taken over the airfield, with the Vienna War Ministry securing a long-term lease and purchasing the existing 33 hangars, marking Wiener Neustadt as a central site for Austro-Hungarian military aviation.

Above: The Etrich I, featured on a graphically enhanced postcard, is depicted during Karl Illner's historic flight on 29 November 1909. The artistic enhancement includes a stylized sun and sunbeams, underscoring the significance of the event. Developed from Igo Etrich's glider project, the aeroplane underwent extensive modifications, including the installation of a 50 hp Clerget engine. Illner successfully piloted the machine in a series of flights that day, with the longest covering a distance of 4-5 kilometers around the entire airfield. *Credit: Christian Reiterer*

Above: The Voisin-built Farman HF.Ibis, an aeroplane of considerable historical significance, was originally purchased by Henri Farman from Gabriel Voisin in 1907. Farman made several key modifications, including the addition of side-curtains, rubberized fabric, and ailerons after witnessing Wilbur Wright's demonstration in Le Mans. These enhancements led to its re-designation as the Farman I-bis. Like many pioneering aeroplanes of the time, the HF.Ibis underwent a series of further modifications, including a temporary conversion to a triplane configuration. As its operational life progressed, the machine, like others of its era, either sustained damage beyond repair or was deemed no longer safe to fly. Eventually, it was sold to the Wiener Syndikat zur Veranstaltung von Schauflügen [Vienna Syndicate for Organizing Demonstration Flights] and played a prominent role in Austria's first aviation exhibition in April 1909, piloted by the French aviator Georges Théophile Legagneux.

Civilian support came from prominent organizations such as the Österreichische Flugtechnische Verein [Austrian Aeronautics Society] and the Österreichische Aëro-Klub [Austrian Aero Club], along with influential individuals like Viktor Silberer. The involvement of the commercial sector was marked by the participation of Eduard Fischer and Ferdinand Porsche from the Austro-Daimler Motor Company, whose expertise and resources were vital.

The municipal leadership, represented by the mayor of Wiener Neustadt, Franz Kammann, played a significant role in facilitating the allocation of land for the airfield. Together, these military, civilian, and commercial interests successfully persuaded the Wiener Neustadt *Stadtgemeinde* [municipal council] to allocate a suitable area for the airfield.

The community built and rented hangars to interested parties, and by 1912, Wiener Neustadt served as both a military and civil airfield. However, in 1912, civil aviation activities moved to Vienna/Aspern, and by 1913, the military took over Wiener Neustadt entirely. The Vienna War Ministry subsequently rented the entire airfield for 25 years and purchased the existing 33 hangars from the Wiener Neustadt community.[3]

Following the decision to open an airfield, various entities were invited to establish themselves at Wiener Neustadt for economic or research purposes.[4] Early activities were primarily civilian, with Ignaz "Igo" Etrich, later known for his *Taube* aeroplane, being the first to relocate his small aeroplane building shop from Vienna to Wiener Neustadt in July 1909. Many others followed suit.

The war ministry had not yet allocated any funding for aeroplanes. However, a group of wealthy Aus-

Above: Georges Théophile Legagneux, a pioneering French aviator, piloted the Voisin-built Farman HF.Ibis during Austria's first aviation exhibition in April 1909. Renowned for his daring flights, Legagneux was a key figure in the early days of aviation.

tro-Hungarian's, known as the Wiener Syndikat zur Veranstaltung von Schauflügen [Vienna Syndicate for Organizing Demonstration Flights], prompted the war ministry to accept aeroplanes in the army. One member, theater director Wilhelm Karczag, purchased the Voisin-built Farman HF.Ibis from France.[5] The aeroplane was transported to Vienna to participate in what would be Austria's inaugural flight exhibition. On 22 and 27 April 1909, near Vienna in the locality known as *Indunationsgebiet*, a flat area of the Danube floodplain, the aeroplane was piloted by the celebrated French aviator Georges Théophile Legagneux.[6]

On each of these days, the aeroplane completed two brief hops, representing the first attempts of an aeroplane to fly in Austro-Hungary. However, the attempts proved unsuccessful, as the aeroplane could not take off, leading to a great deal of anger among the spectators. Legagneux had to be brought to safety by the police from the angry crowd. Following these events, the aeroplane was donated to the *k.u.k. Armée*, where it became the first aeroplane in the inventory of the *k.u.k.* Military Aviation Institute. Intending to enhance its flight performance, a series of modifications were initiated. Despite these attempts at improvement, the aeroplane ultimately failed to fly, prompting its donation to the *Heeresgeschichtliches Museum* [Military History Museum] later that year.[7]

On 15 August 1909, the *k.u.k. Heeresverwaltung* [Army Administration] rented two of the hangars in Wiener Neustadt and declared them to be a *Militär aviatische Station* [military aviation station] and selected *Hptm* Wilhelm Booms to oversee the responsibilities. Donations to the military were not unusual in the *k.u.k.* Monarchy, with local communities or businesses occasionally financing barracks construction and donating them to the war ministry to attract military units to their town.[8]

Consequently, the war ministry accepted the donation of the aeroplane and rented two hangars in Wiener Neustadt for the disabled Farman, hoping to repair it and use it for pilot training.[9] However, the Farman never reached its potential.[10] Austria-Hungary was home to innovative engineers, and during this period, the civil sector engaged in spirited discussions about new aviation developments, like other major European states. Civilian Austro-Hungarian aviators and young army officers attracted attention; however, this did not significantly sway Vienna War Ministry decision-makers, who persisted in their cautious approach. Many high-ranking officers regarded flying an aeroplane as mere frivolity.[11]

As often occurred, developments within the civilian sphere catalyzed military decisions. *Genmjr* Leopold Schleyer, who headed the Troop Transport Brigade, which included the proposed the establishment of a third branch within this department to focus specifically on aeroplanes, in addition to the existing balloon and airship divisions. Although the war ministry showed some interest, no further action was taken.[12]

On 19 October 1909, a second military aeroplane, a Wright biplane, arrived in Wiener Neustadt. It was provided for military use by Vienna industrialist Baron Robert von Lieben. Before it could be used for training military pilots, it was decided to modify the machine. This was carried out under the guidance of Richard Knoller. The skids were removed, and a four-wheel undercarriage was installed obviating the need for the catapult launch ramp. Additionally, he made changes to the biplane front elevator section to a monoplane. The success of both modifications was adopted by the

Above: Voisin-built Farman HF.Ibis in its biplane configuration on the field in Austria, with Georges Théophile Legagneux at the controls during the flight exhibition at the *Induationsgebiet* (floodplain), May 1909. *Credit: Christian Reiterer*

German firm building Wrights under license.[13]

Another influential event transpired when the renowned English Channel-crossing pilot Louis Blériot visited Budapest and Vienna in late October 1909 at the invitation of Hungarian entrepreneur Charles Müller and Austrian-born Alfred Ritter von Pischof, commercial director of the Blériot aeroplane factory in Paris. On 23 October 1909, an enthralled crowd of 200,000–250,000 people, including Emperor Franz Joseph I, members of the Royal and Imperial Habsburg Family, and high-ranking military officers, witnessed an aeroplane in flight for the first time at Simmeringer Heide. Known for its open fields and frequent use for public and military events, Simmeringer Heide was situated on the outskirts of Vienna yet within the city's administrative boundaries. This moment marked the seemingly unstoppable rise of aeroplanes.[14]

On 17 November 1909, the Wiener Neustadt site was officially designated as an airfield by the civilian *Österreichische Aeronautische Kommission* [Austrian Aeronautics Commission], and the two military hangars and their small ground support crew of soldiers were placed under the command of *k.u.k. Hptm* Wilhelm Booms.[15] This marked the birth of *k.u.k.* aviation in the town.

Due to the dual nature of the *k.u.k.* state, which consisted of Austrian and Hungarian halves, each with independent domestic representations, political symmetry necessitated the establishment of a military aerodrome at Rákos-mező near Budapest when Wiener Neustadt was declared a military airfield.[16] Nevertheless, Rákosmező never achieved the same level of significance as Wiener Neustadt.

Austrian industrialists and technicians were not idle during this time. Engineering professor Richard Knoller modified the Wright biplane on behalf of Baron von Lieben. Knoller, who held the first chair for *Luftschifffahrt und Automobilwesen an der Technischen Hochschule Wien* [Aeronautics and Automotive Engineering at the Vienna Polytechnic Institute], is considered the founder of aeronautics in Austria-Hungary and later became known as an aeroplane designer.[17] Numerous early scientists, engineers, and industrialists working in Austro-Hungary have been forgotten, but a few notable names include Ferdinand Porsche, Igo Etrich, Richard Knoller, Alfred Pischof, Ernst Körting Jr., Ludwig

Above: *Flugfeld* [Flying Field] Wiener Neustadt, depicted here in 1910, with its early hangars and infrastructure supporting both military and civilian aviation activities. On 15 August 1909, the *k.u.k. Heeresverwaltung* [Imperial and Royal Army Administration] rented two of these hangars, officially establishing a '*Militär aviatische Station*' [Military Aviation Station]. This marked a significant milestone in Austro-Hungarian military aviation, as the field transitioned into a strategic site for pilot training, aeroplane testing, and operational development under the direction of *Hptm* Wilhelm Booms. The station would play a pivotal role in shaping the empire's aviation capabilities during this formative period.

Lohner, Josef Sablatnig, Adolf Warchalowski, and Karl Paulal.[18]

The prevailing antagonistic attitudes of high-ranking *k.u.k.* military officers toward flying activities were exemplified by an incident involving pioneering aviator *Hptm* Emanuel Quoika of *Feldkanonen-Regiment 6* [Field Artillery Regiment 6] and the *Gen.-Stabschef des II. Korps* [Chief of Staff of *k.u.k.* II. Corps] Vienna.

After learning to fly in France, Quoika purchased a Blériot XI (№. 22) aeroplane in November 1909 with his own money. With the acquiescence of his superiors, he built a wooden hangar on the grounds at Simmeringer Heide to house the machine. Quoika began flying his Blériot monoplane from the area, which was situated on the outskirts of Vienna yet within the city's administrative boundaries.

Upon learning of these flights, the *Gen.-Stabschef des 2. Korps* [Chief of Staff 2. Corps] called Quoika to report to him. The *Oberst* [Colonel] questioned whether Quoika wished to continue his aerial acrobatics nonsense or fulfill his duty as an artillery commander, threatening to send an engineer squad to demolish the hangar if Quoika did not cease his ridiculous flying activities. With no other option, Quoika complied. Similar episodes involving pioneering pilot officers were not uncommon elsewhere in Europe at the time.[19]

Although limited to the military environment, numerous private sportsmen, entrepreneurs, civilian enthusiasts, and both well-established and newly founded industrial companies cautiously ventured into the emerging aeroplane business by renting hangars in Wiener Neustadt. The decision to locate in this area was primarily due to the presence of the Oesterreichische Daimler-Motoren-Aktiengesellschaft [Austro-Daimler Corporation]. Engines, alongside airframes, would prove to be an equally crucial component particularly for pioneer period aeroplanes. In Wiener Neustadt, engineers and aeroplane designers had easy

Above: Josef Sablatnig flying a modified Wright biplane at Wiener Neustadt Flugfeld in 1909. The modifications, designed by Richard Knoller, included a four-wheel undercarriage to improve ground handling and an altered front elevator to enhance flight stability and control. Provided by industrialist Baron Robert von Lieben, this aeroplane became a pivotal asset for Austrian aviation training and experimentation. Sablatnig, who would go on to become one of Austria's most influential aviators, trained on this aeroplane and performed a series of demonstration flights across Europe. These flights not only highlighted the transformative potential of powered flight but also served as the launchpad for Sablatnig's distinguished aviation career.

access to a leading aeroplane engine producer.[20] Notably, the Austro-Daimler Corporation originated as an affiliate of the German *Daimler Motoren Gesellschaft* [Daimler Engine Company] but became independent from its former German parent company in 1909.

During the initial months, military engagement at Wiener Neustadt remained limited, with most activities stemming from the civilian sector. In spring 1910, the war ministry received two more aeroplanes: a Blériot monoplane donated by Klemens Maria von Radowitz and a fully airworthy Voisin biplane provided on a temporary basis by the wealthy Imperial Counselor Josef Flesch. Surprisingly, *General der Infanterie(GdI)* [General of Infantry] Franz Xaver von Schönaich, Minister of War, relied heavily on the generosity of Vienna's elite society. While the initial fleet of *k.u.k. Armée* aeroplanes was largely acquired through donations, there were significant contributions made through other means as well.[21]

Endnotes

1. Jäger, *Erkundung*, 78; Stahl, *Die Militärluftfahrt.*, 120; Cuneo, *Winged Mars I*, 89; Morrow, *German Airpower*, 29–33.
2. Peter, *Luftschiffer*, 52.
3. Desoye, 20, 23, 25; Peters, *Militärluftfahrt*, 56; Zelezny, *Neustadt*, 13–15.
4. Peter, *Luftschiffer*, 55.
5. Reg Winstone, *The Right Flyer: Gabriel Voisin, Henry Farman and the Archetype of Aeroplanes*, ISBN-

Above: Broadsheet poster advertising Louis Blériot's demonstration flight at Simmeringer Heide, Vienna, on 23 October 1909. The event, attended by Emperor Franz Joseph I, members of the Habsburg family, and high-ranking military officials, marked the first time many in Austria witnessed an aeroplane in flight. The poster emphasizes the emperor's arrival, the scheduled flights, and a military band concert during the event. With an estimated 200,000 to 250,000 spectators present, this event underscored the growing fascination with aviation in Austria-Hungary.

10:0956981100, 231–236; Henry Farman modified this Voisin-built biplane machine to a triplane configuration on 16 November 1908; then on 28 November it was converted back to a biplane. In January 1909, before being sold to Wiener Syndikat zur Veranstaltung von Schauflügen [Vienna Syndicate for the Organization of Flight Demonstrations] it was returned to a triplane arrangement; *French Aeroplanes before the Great War*, Leonard E. Opdycke, 118.

6. "La vie sportive: aéronautique," *Le Matin*, ⊠. 9193, 29 avril 1909, 4; A series of public flights were made on 21, 22, and 23 May with limited success. Winstone, *The Right Flyer*, 231–236.
7. Winstone, *The Right Flyer*, 231-236; Hugo Kerchnawe, *Die unzureichende Kriegsrüstung der Mittelmächte als Hauptursache ihrer Niederlage. Ergänzungsheft 4 zum Werke 'Österreich-Ungarns Letzter Krieg'.* [The Central Powers' inadequate armament for war as the main cause of their defeat. Supplement 4 to the work "Austria-Hungary's Last War"] *Militärwissenschaftlichen Mitteilungen*. Wien 1932, 38; Peter, 52; Gerhard Artl, *Die Entwicklung*, 2.
8. Manfried Rauchensteiner, *The First World War and the End of the Habsburg Monarchy, 1914-1918,* Vienna: Böhlau, 2014, 56.
9. Peter, *Luftschiffer*, 52.
10. Keimel, *Österreichs*, 86.
11. Peter, *Luftschiffer*, 52, 59.
12. Desoye, 17; Peter, *Luftschiffer*, 56, 64.
13. Peter, *Luftschiffer*, 56; Zelezny, *Neustadt*, 19, 172. The first flight was on 4 January 1910 by Vincent Wiesenbach from Luxembourg.
14. *Vasárnapi Újság* 43. Sz. 1909. (56. Évfolyam.) Budapest, október 24. [Sunday newspaper ⊠. 43 1909 (Volume 57), Budapest, 24 October], 898-902; Peter, *Luftschiffer*, 58. The crowd size represented approximately one tenth to one eighth of Vienna's population at the time.
15. Gerhard Artl, *Die Entwicklung der k.u.k. Luftfahrtruppen 1909-1918*, 2, [The development of the *k.u.k.* air forces 1909–1918] Lecture, Österreichische Verkehrswissenschaftliche Gesellschaft, 14 October 2014; Marcus F. Zelezny, *Die ersten Passagierflüge, erster hoher Besuch underste aviatische Wettbewerbe 1910 in Österreich* [The first passenger flights, first royal visit, and first aviation competitions in 1910 in Austria] 4, Österreichischen Flugzeug Historiker [Austrian Aircraft Historians] öfh Nachrichten 1/20.
16. Peter, *Luftschiffer*, 56.; The origins of Hungarian aviation can be traced back to the summer of 1909 when the construction of hangars commenced at Rákos-mező [Rákos Field] for the purpose of assembling and storing aeroplanes. The first hangar was owned by Ágoston Kutassy, a pioneer in Hungarian aviation who also became the first member of the *Magyar Aero Club* [Hungarian Aero Club] to be awarded a pilot's certificate, this after his flight of on 22 December 1910 in his Farman HF.III. By 1912, the number of hangars at the site had increased to 17. These hangar structures were primarily built using wooden planks and featured roller sliding doors that operated on tracks. With the exception of the initial two hangars, each of these structures spanned 20 meters in width, 12 meters in depth, and 6 meters in height.
17. Hermann Hoernes, *Buch Des Fluges, III. Band.* [Book of flight.III.Band.], Wien 1912. Verlag Georg Szelinski, K.K. Universitäts-Buchhandlung. 316–317.
18. (in addition) Theodore von Karman, Wolfgang Benjamin Klemperer, István Petróczy von Petrócz , Julius von Berg, Wilhelm Zurovec, Oszkár Asboth, Josef Mickl, Franz Xaver Wels, Anatol & Alexander Renner, Wilhelm Kress.

Above: Louis Blériot preparing for his historic 1909 flight at Simmeringer Heide, Vienna. On 23 October 1909, in front of a massive crowd that included Emperor Franz Joseph I and the Habsburg royal family, Blériot demonstrated the remarkable capabilities of his Blériot XI aeroplane, marking the first time many Austrians witnessed an aeroplane in flight. This event played a key role in igniting public and military interest in aviation within Austria-Hungary. *Credit: Christian Reiterer*

19. Peter, *Luftschiffer,* 55–58; In the early years of aviation, prior to 1914, most countries, including France, Germany, England, Italy, and Austria-Hungary, were still formalizing their military aviation programs, and private aeroplane ownership among military personnel was uncommon. These countries imposed several restrictions on military officers who wished to fly or own private aeroplanes. These officers needed permission from their commanding officers to own or operate private aeroplanes and were limited in the types of aeroplanes they could fly, as well as the geographic areas over which they could operate. Specifically, military pilots were subject to stringent restrictions when participating in air meets. These measures were implemented to ensure safety, maintain discipline, and protect national security. Starting in 1912, military pilots were officially prohibited from earning additional income from flying, leading many to participate under pseudonyms. This practice was widely adopted across several countries, reflecting the tension between emerging aviation enthusiasm and military regulations.
20. Michael Düsing, *German & Austro-Hungarian Aero Engines of WWI: Vol. 1*, Aeronaut Books, 2022, 88–112.
21. Peter, *Luftschiffer,* 56; Keimel, *Österreichs,* 298.

Above: Österreichische Daimler-Motoren Aktiengesellschaft (Austro-Daimler) 60 hp Engine, designed by Ferdinand Porsche, was a pioneering 4-cylinder inline aeroplane engine produced in 1910. This engine featured a 6.3-liter displacement with a bore of 120 mm and a stroke of 140 mm. It was capable of delivering a maximum output of 65 hp at 1350 RPM. The engine utilized a Daimler rotary slide carburetor and a Bosch high-tension magneto ignition system, making it a robust choice for early aviation. Its design was notable for its water-cooled system, ensuring reliability in various flight conditions.

5. Manufacturing Aeroplanes

Above: The Lohner-Daimler *Pfeilflieger* [Arrow Flyer] prototype, designed by Karl Paulal, with *Rittmeister* Hans *Ritter* Umlauff von Frankwell standing in front. This innovative design featured distinctive swept-back wings, with a 15 percent wing sweep contributing to its aerodynamic profile and inspiring the name *Pfeilflieger*. The upper wing spanned 13.5 meters, while the lower wing measured 9.2 meters, paired with a 9.7-meter-long fuselage. Powered by a four-cylinder 65 hp Austro-Daimler aero-engine driving a 2.7-meter Lohner propeller, the aeroplane embodied early attempts to combine speed and stability. Umlauff's test flights were instrumental in refining its operational capabilities. Despite economic challenges Lohner Wagenfabrik established itself as a key supplier of military aeroplanes for the *k.u.k. Luftschifferabteilung*, marking the start of its significant contributions to Austro-Hungarian military aviation. *Credit: Christian Reiterer.*

Unlike the military bureaucrats in the Vienna War Ministry, industrialists were far-sighted. Pioneers and forerunners of the later *k.u.k.* aeroplane industry, such as Igo Etrich, designer of the first successful *k.u.k.* military aeroplane the Etrich *Taube*; Jacob Lohner & Co. Hof-Wagenfabrik; the engine factory Werner & Pfleiderer; Motor-Luftfahrzeug–Gesellschaft; Österreichisch-Ungarische Autoplan, and others rented hangars and established workshops in Wiener Neustadt.[1]

Of the many newcomers and established companies in Wiener Neustadt, only a few survived financially. Notably, Lohner Wagenfabrik later became the primary supplier of *k.u.k.* military aeroplanes. The factory, managed by Ludwig Lohner, saw its chief engineer Karl Paulal roll out the first prototype of the famous *Pfeilflieger* [Arrow-flyer] for its maiden flight on 3 November 1910. Paulal would later develop all Lohner *Pfeilflieger* versions until 1916, when he was superseded by Leopold Bauer.

Months prior, in April 1910, Igo Etrich's *Taube II*, piloted by Karl Illner, his chief mechanic and pilot, completed the aeroplane's maiden flight at Wiener Neustadt.[2] Initially underpowered, the *Taube*'s

Above: The Lohner *Pfeilflieger*, a cornerstone in the development of Austro-Hungarian military aviation. Powered by an Austro-Daimler engine, this aeroplane became instrumental in advancing reconnaissance techniques and programs for the Austro-Hungarian military. Karl Paulal's engineering expertise drove the *Pfeilflieger's* evolution, with subsequent models incorporating more powerful engines and refined aerodynamics. Despite delays in the procurement process for aeroplanes for military use—manufactured at the Wiener Neustadt facility—Lohner Wagenfabrik firmly established itself as a leading supplier of aeroplanes to the *k.u.k. Luftschifferabteilung* during this critical early period.

success was catalyzed by the incorporation of the Austro-Daimler 65-hp engine. The aeroplane's ease of operation, remarkable flight stability, and forgiving speed range contributed to the development of various subsequent iterations. The *Taube IV*, powered by the 65-hp Austro-Daimler engine, achieved commercial success.

Etrich's establishment in Wiener Neustadt focused on research and development rather than aeroplane production. Consequently, Etrich collaborated with Lohner's factory for production and distribution of the first *Taube* series. This partnership would not be the sole production arrangement for the *Taube*.

A savvy businessman, Etrich's interests encompassed both aeroplane development and profit generation. He sold the *Taube* patents to the newly formed Motor-Luftfahrzeug-Gesellschaft, a joint venture between Oesterreichische Daimler-Motoren-Aktiengesellschaft and Lohner-Werke. The Motor-Luftfahrzeug-Gesellschaft primarily functioned as a sales organization for Lohner-produced *Taube* aeroplane.

Etrich also remained active in Wiener Neustadt, establishing a flight school for Austrian and German pilots. It did not take long for Vienna War Ministry to begin manufacturing the *Taube* at the Fischamend *Luftfahrtabteilung* [Aeronautics Department]; in doing so, they (illegally) obviated paying licensing fees.

Furthermore, Etrich licensed the German E. Rumpler-Flugzeugbau GmbH. However, the German patent office invalidated Etrich's licenses, asserting that they encompassed publicly accessible knowledge and expertise. This decision led to unregulated *Taube* production in Germany and the emergence of various models, such as the Rumpler *Taube*, Albatros *Taube*, and Jeannin *Stahletauben*.

The Taube ultimately became the most successful aeroplane of the pioneer prewar era due in large part to its self-stable flight tendencies. Employed as a

Above: Igo Etrich's *Taube II*, named for its dove-like wing shape, took its maiden flight at Wiener Neustadt in April 1910, piloted by Karl Illner. During that month, a series of test flights showcased the aeroplane's impressive capabilities, advancing from brief field hops to extended journeys lasting up to an hour and covering 84 kilometers. On 17 May 1910, the *Taube II* achieved its first cross-country flight, successfully landing at Simmeringer Heide before returning to Wiener Neustadt. These milestones underscored its significance as a groundbreaking and dependable advancement in early aviation.

trainer in flight schools, a military reconnaissance aeroplane, and a sports aeroplane for private pilots, the Taube experienced no severe accidents throughout 1910. By 1915, when the aeroplane became outdated, approximately 500 *Taube* planes had been constructed worldwide.[3]

Concurrently, Adolf Warchalowski of Werner & Pfleiderer's company garnered significant attention with his Type Autoplan *Biplan*, the first of which was named *Vindobona*. Adolf Warchalowski, a lesser-known yet influential figure during the pioneer period in Wiener Neustadt, was an engineer employed by the Werner & Pfleiderer company, a producer of technical equipment and engines. Recognizing the potential of the burgeoning aviation industry, the Werner & Pfleiderer directorate dispatched Warchalowski to Paris to acquire flight training and procure a Farman biplane. In November 1909, the Farman biplane made its first hop in Wiener Neustadt, with full flight operations commencing in February 1910. Warchalowski subsequently became the first Austro-Hungarian pilot to be licensed by the newly established national Österreichische Aero-Club on 22 April 1910.[4] This organization was responsible for overseeing and regulating flight training and pilot licensing in the nation.

Werner & Pfleiderer's acquisition of the Farman biplane was primarily aimed at replicating and refining an existing design. Warchalowski was assigned the task of enhancing the biplane, and by May 1910, a mere six months after the arrival of the French aeroplane in Wiener Neustadt, he had successfully modified and developed the Autoplan *Biplan*, an improved machine. The aeroplane was initially powered by a 50-hp Gnôme engine but was later upgraded to a 65/70-hp Daimler four-cylinder inline water-cooled engine. Warchalowski's machine performed exceptionally well, earning him numerous accolades in national competitions. His piloting expertise was even trusted by high-ranking members of the Royal Habsburg

Above: The Etrich *Taube II* at Wiener Neustadt, equipped with visible wingtip wheels, was designed to minimize the risk of ground loops—a dangerous condition in which an aeroplane pivots uncontrollably during landing. These wheels not only protected the wingtips but also significantly reduced the chances of major or catastrophic accidents during landings on the uneven, unpaved airfields common at the time. This innovation contributed to safer landings and represented an important step forward in aviation safety. *Credit: Christian Reiterer*

Family, who frequently accompanied him on passenger flights. Additionally, *Chef des Generalstabs* Conrad von Hötzendorf had Warchalowski conduct an orientation flight in the aeroplane at Wiener Neustadt on 15 September 1910.[5]

In an effort to produce both Warchalowski's and Pischof's aeroplanes, Werner & Pfleiderer, alongside Austrian investors including August Warchalowski (Adolf's elder brother) and Hungarian investor Oskar von Wahl, established the Österreichisch-Ungarische Autoplan-Werke GmbH in February 1911. Their aim was to supply aeroplanes to the military. Unfortunately, the venture was unsuccessful, leading to the company's eventual bankruptcy in 1914.[6]

Following his orientation flight with Adolf Warchalowski on 15 September 1910, *Chef des Generalstabs* Conrad submitted an application to the war minister on 5 October 1910 requesting the establishment and provisioning of an aviation force comprising 200 aeroplanes and 400 pilots in total. He envisioned aviation units assigned to each army corps, division, and fortress. However, the war minister did not fully support von Hötzendorf's proposal. Instead of actively pursuing the creation of an army aviation service, the ministry purchased a single Blériot monoplane and issued a tender outlining technical requirements for future military aeroplanes.[7]

During 1910, Wiener Neustadt became a hub of aviation activity. Numerous professional and private Austro-Hungarian designers, officers, and companies involved in vehicle or engine production engaged in aviation pursuits in the city. Though many of these entities have since been forgotten, 13 new aeroplane designs were presented or tested in Wiener Neustadt in 1910 alone. Companies such as Austro-Daimler, Lohner, Werner & Pfleiderer, and several lesser-known firms, ventured into the aviation engine business. Werner & Pfleiderer, for example, manufactured Hiero engines, named after their designer, German-born motorcyclist Otto Hieronimus.[8]

In 1910, Austro-Daimler achieved notable success with their 65-hp four-cylinder, 105 kg aeroplane

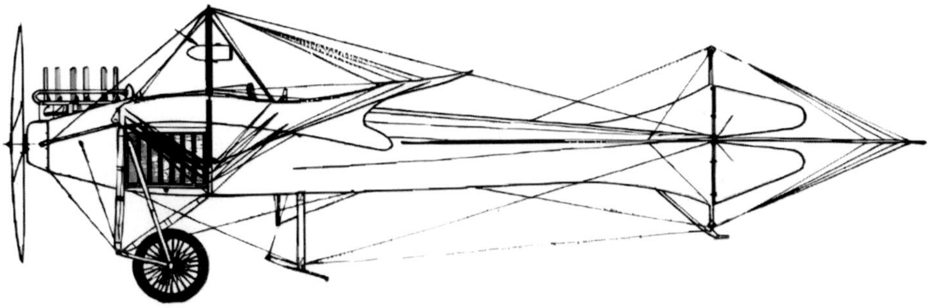

This Austro-Hungarian Etrich Taube, powered by a 6-cylinder Austro-Daimler engine, exemplifies the design principles of early aviation. Renowned for its exceptional flight stability, the Taube became a favored reconnaissance aeroplane. Early versions relied on wing warping for lateral control, achieved by twisting the wings to manage roll. Pitch control was provided by warping the rear half of the stabilizer, functioning as an elevator, while yaw control was managed through hinged, twin triangular rudder surfaces. Its distinctive bird-like wings were inspired by the gliding seeds of the Alsomitra macrocarpa plant. Licensed for production by Lohner-Werke in Austria and Edmund Rumpler in Germany.

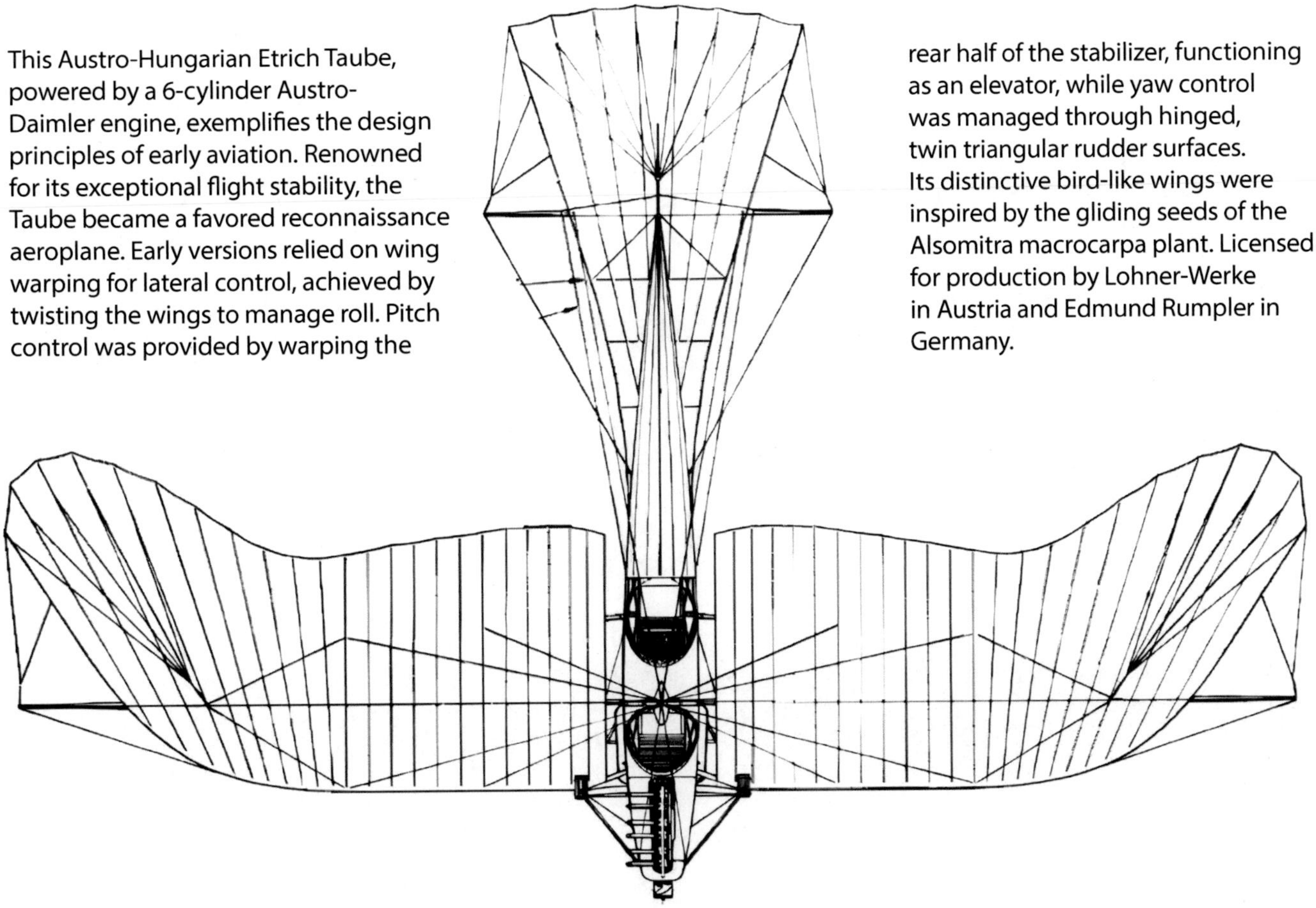

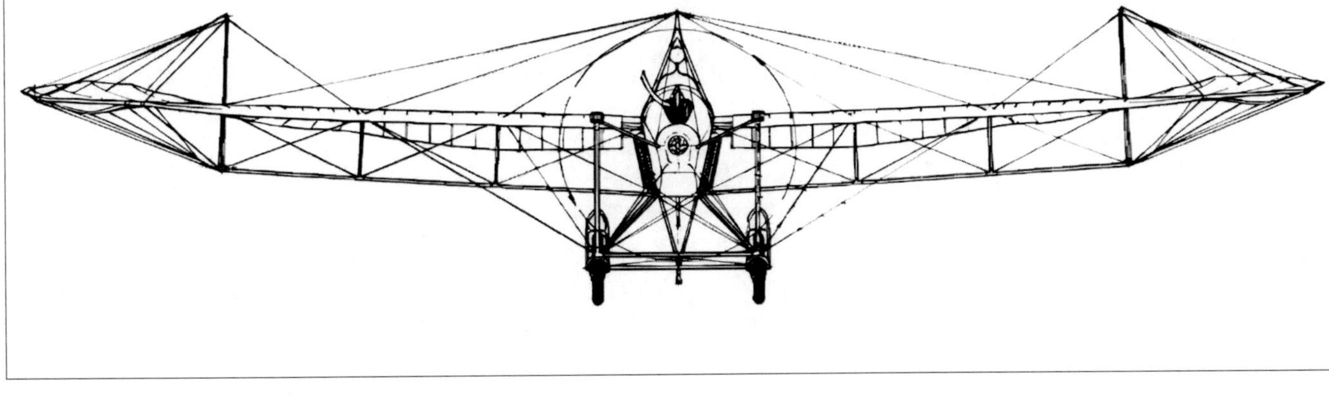

Left: Advertisement for Motor-Luftfahrzeug Gesellschaft (MLG) promoting the Original Etrich *Taube* and Lohner *Pfeilflieger*. This image captures the early ambitions of Austria-Hungary's aviation industry to establish a national identity in aeroplane manufacturing. The MLG, a joint venture between Austro-Daimler and Lohner-Werke, played a key role in marketing and distributing aeroplanes like the Etrich *Taube*, a highly influential design at the time. Despite the initial success of these models, limitations in production scale and engineering challenges prevented the company from fully meeting the demands of the *k.u.k.* military prior to World War I.

engine, designed by Ferdinand Porsche. The engine's performance was increased to 90–120 hp within a year, marking the beginning of Austro-Daimler's ongoing business in aeroplane engine development and production. The reputation of the 65-hp four-cylinder engine rested on its exceptional power-to-weight performance and outstanding reliability.[9] However, it is important to note that during the early months of the war, both the Austro-Daimler and Hiero engines faced reliability issues due to factors such as design flaws, manufacturing issues, material shortages, and maintenance or operational problems, all of which were exacerbated by the pressures of wartime production and use.[10]

Endnotes

1. Erich Kahlen, *Die Entwicklung der österreichisch-ungarischen Fliegertruppe von 1914-1918.* [The development

Above: Karl Illner's successful cross-country flight in an Etrich *Taube* from Vienna/Simmeringer Haide to Horn and back on 10 October 1910. His landing at Horn, with Ferdinand Porsche standing to the left of the propeller and pilot Karl Illner to the right, completing the 150-kilometer round trip earned him a prize of 20,000 kronen. *Credit: Christian Reiterer*

of the Austro-Hungarian Air Force from 1914–1918] Kriegsarchiv Wien (Luftarchiv). MS/Lu8, 52.

2. Karl Illner, a seminal figure in early Austro-Hungarian aviation, served dual roles as chief mechanic and lead pilot for Igo Etrich. Born in Schatzlar, East Bohemia, Illner was pivotal in the early testing and development of Etrich's aeroplanes, particularly the Etrich *Taube* series. His responsibilities spanned the intricate mechanical work necessary to prepare aeroplanes for flight and the adept piloting required for their operation. Illner's technical expertise was essential in overcoming the initial challenges faced by the *Taube*, particularly issues with power and stability. His hands-on approach involved not only routine adjustments and maintenance but also significant adaptations that improved aeroplane performance over time. This included collaborating with manufacturers like Austro-Daimler to enhance engine reliability and efficiency, critical for the *Taube*'s operational success. Beyond his mechanical acumen, Illner's leadership in conducting flight tests and his ability to train other pilots and mechanics significantly contributed to the development of the *Taube* and the advancement of aviation technology in Austria-Hungary.
3. Zelezny, *Neustadt*, 215-216; The Etrich *Taube*, due to its unique design, was an inherently stable aeroplane. After a disturbance caused by a gust of wind, it typically resumed its original flight attitude without pilot intervention. Karl Illner repeatedly demonstrated this stability by briefly standing up in the aeroplane and waving to spectators with both hands, while the plane continued to fly straight ahead, highlighting the aeroplane's ability to maintain stable flight without constant control inputs.
4. *K.u.k. Österreichischen Flugtechnischen Vereines 1913*, [*K.u.k.* Austrian Aeronautics Association 1913], 168; The Österreichischer Aero-Club, originally established as the Wiener Aëro-Klub on 7 March 1901 and renamed in 1910, becoming the Österreichischer Aero-Club, was responsible for certification and regulation of flight training and pilot licensing in Austria. It was affiliated with the Fédération Aéronautique Internationale (FAI) on 27 May 1910. In Hungary, the Magyar Aero Club, founded on 26 February 1910, independently oversaw similar responsibilities. Both organizations played key roles in advancing aviation standards and practices within their respective regions, joining the FAI to gain international recognition and compliance.
5. Gerhard Artl, *Die Entwicklung*, 4.
6. Karl Tumlirz, *Österreichisches Biographisches Lexikon 1815–1950 Band XV: Warchalowski, August* [Austrian biographical lexicon 1815-1950, volume XV: Warchalowski, August], 487; Zelezny, *Neustadt*, 208, 210, 214; Overall, they sold only one aeroplane to the military, the Autoplan *Hydroplan*, failed to sell the Autoplan military biplane, and did not sell any of the license-

Above: Etrich Type VII *Renntaube*: Culmination of Eight Record-Breaking Achievements—On 4 October 1911, during the *Österreichische Flugwoche* [Austrian Aviation Week], the *Renntaube*, powered by a 120 hp Austro-Daimler engine, completed a historic 112 km flight from Vienna with two passengers. This photograph captures their triumphant landing and features (from left to right): 1. *Oblt* Eduard Babouschek (passenger), 2. Lilly Steinschneider (passenger and Hungary's future first female pilot), 3. *Oblt* Heinrich Bier (pilot), 4. Heinrich Leszuk (*Baumeister* [chief builder]), 5. Ferdinand Porsche (automotive engineer), and 6. Camillo Castiglioni (industrialist). Earlier that week, Bier had already set seven world records in speed, endurance, and altitude flights in the same aeroplane, solidifying his status as a pioneering aviator.

built Nieuport monoplanes. Consequently, they went bankrupt; Österreichische Industrie-Werke Werner & Pfleiderer, Kommandit-Gesellschaft, initially founded on 20 December 1910, focusing on aeroplanes such as the Pischof *Monoplan* and Autoplan *Biplan*, along with early Hiero aero engines. It transformed into Österreichisch-Ungarische Autoplan-Werke GmbH on 14 February 1911 with shareholders like August Warchalowski (Vienna) and Oskar von Wahl (Budapest). This entity expanded its range to include the Autoplan *Monoplan*, *Biplan*, *Hydroplan*, and *Militär Doppeldecker* but went into liquidation on 15 May 1914. Subsequently re-established on 7 August 1914 as Österreichische Industriewerke Warchalowski, Eissler & Co GmbH, it was spearheaded by August Warchalowski and Georg Eissler, focusing on Hiero aero engines until its deregistration on 29 December 1916. On the same day, it was reconstituted as Österreichische Industriewerke Warchalowski, Eissler & Co AG, adding shareholders like Richard Werner, Otto Werner, and Ernst Schwarz, and continued until its final deregistration on 14 December 1928.

7. Peter, *Luftschiffer,* 69.
8. Düsing, *Vol. 3*, 207-212.
9. Karl Meindl, *Luftsiege der k.u.k. Luftfahrtruppen. Russische Front 1914-1918.* [Aerial victories of the *k.u.k.* aviation groups. Russian Front 1914–1918.] öfh Nachrichten. Sonderheft 28. Wien 2001, 4, 465.; Düsing, Vol. 1, 88-112; Zelezny, *Neustadt*, 198.
10. Meindl, *Luftsiege*, 4.

Above: Adolf Warchalowski's Endurance Prize Flights—On 19 February 1910, in front of over 1,000 spectators at the Wiener Neustadt aerodrome, Adolf Warchalowski achieved a significant milestone with an endurance flight lasting over 16 minutes. This accomplishment earned him his first win in the prestigious Die Gerngroß-Preise, established by Alfred and Hugo Gerngroß, prominent Viennese business owners. The prizes were instituted to advance Austro-Hungarian aviation, with a key stipulation that competitors must be Austrian pilots operating their own aeroplane. The awards offered 2,000 kronen for a flight exceeding 15 minutes, 4,000 kronen for a five-minute flight with a passenger, and 10,000 kronen for a 30-kilometer cross-country flight from one location to another and back. Flying the Farman owned by the Werner & Pfleiderer company, Warchalowski also carried *Hptm* Wilhelm Booms as a passenger on a subsequent flight that same day, accomplishing two of the three prize flights and cementing his role as a pioneer in Austro-Hungarian aviation.

Below: Adolf Warchalowski at the controls of his modified Farman III biplane, a machine formerly owned by Werner & Pfleiderer but now fully his, achieved a historic moment on 12 March 1910 at Wiener Neustadt. Accompanying him was his sister-in-law, Anna Warchalowski, who became the first female passenger in Austria-Hungary. This flight held added significance, as Warchalowski's previous attempts to claim two of the Die Gerngroß-Preise had been disqualified due to a technicality—at the time, he did not legally "own" the aeroplane. Now, as its rightful owner, he effortlessly repeated both prize-winning flights, this time with Anna as his passenger.

Above: Alfred *Ritter* von Pischof's *'Monoplan,'* photographed at Wiener Neustadt in 1911. Designed by von Pischof and produced in series by the Österreichisch-Ungarischen Autoplan Werke in Vienna from 1910 to 1914, this early monoplane embodied innovation and practicality. Its design demonstrated remarkable stability, capable of carrying a passenger. Flight accomplishments included Austria's first overland flight on 9 May 1910, when von Pischof piloted the aeroplane with his wife Sophie as a passenger on a cross-country flight of 45 kilometers in 14 minutes, covering a route between Wiener Neustadt and Seebenstein before returning to the airfield. Further trials in early 1910 showcased its reliability, with 18 test flights conducted over three weeks, including seven passenger flights with Hptm Wilhelm Booms and Josef Pichler, the operations manager at Werner & Pfleiderer. It featured a 65 hp Austro-Daimler engine and a lightweight yet durable frame, with specifications such as a 27 m^2 wing area and a maximum takeoff weight of 550 kg. Its top speed of 85 km/h made it competitive in the burgeoning field of aviation.

Left: *General der Infanterie Freiherr* Franz Conrad von Hötzendorf, Chief of the General Staff of the Austro-Hungarian Army, was an early advocate for integrating aviation into military operations. Despite his strong support for developing an aviation force, the Austro-Hungarian War Ministry was hesitant to fully embrace his vision, resulting in a slower pace of early military aviation development.

Above: At the controls of the original Lohner-Daimler *Pfeilflieger* is *Rittmeister* Hans *Ritter* Umlauff von Frankwell, a pioneering aviator, army officer, and graduate engineer. Umlauff played a crucial role in the development of Lohner's early designs, sharing his experiential knowledge in aeroplane construction with Karl Paulal. This expertise was drawn from his earlier glider work and the two aeroplanes built for him by Lohner. Powered by a 65 hp Austro-Daimler engine designed by Ferdinand Porsche, the *Pfeilflieger* made its maiden flight on 18 October 1910. Umlauff gained fame piloting and winning in a *Pfeilflieger* during the prestigious Vienna-Budapest-Vienna air race on 24 June 1911. In 1913, he became the first recipient of the newly established *Feldpilotenabzeichen* [Field Pilot's Badge], marking his prominence in Austro-Hungarian aviation. He was also the third military officer and the tenth individual in Austria-Hungary to earn a pilot certificate. Umlauff would later serve as one of the first members of the *k.u.k. Luftschifferabteilung* and subsequently as a member of the *k.u.k. Luftfahrtruppen*, where his reconnaissance efforts and leadership significantly contributed to the early military use of aviation.

Right: On 25 September 1911, pilot Josef Sablatnig accomplished a remarkable flight with four passengers aboard an Autoplan *Biplan*. Seated to Sablatnig's left is Lilly Steinschneider; behind her is Albert Hutter; to his right is Franz Seidl; and partially hidden behind the control yoke is Hermann Hold. Sablatnig's achievement demonstrated the growing capability of early aeroplanes to carry multiple passengers, marking a significant step in aviation's progression during the pre-war period.

Above: Standing in front of the Lohner aeroplane, from left to right, are Ferdinand Porsche, Karl Illner, and Karl Paulal. These three figures were pivotal in shaping Austro-Hungarian aviation. Ferdinand Porsche, a visionary engineer, made significant contributions to engine development, particularly with the Austro-Daimler engines. Karl Illner, chief mechanic and pilot for Etrich, conducted numerous test flights, establishing himself as a key figure in early Austro-Hungarian aviation. Karl Paulal, Lohner's chief engineer, was instrumental in developing military aeroplanes for the *Luftschifferabteilung*, including the renowned *Pfeilflieger*, positioning Lohner at the forefront of the Austro-Hungarian aviation industry. Together, they propelled the success of Motor-Luftfahrzeug-Gesellschaft's (MLG) efforts to advance aviation in the region.

6. The Air Meets

Above: Broadsheet poster of the Internationales Wettfliegen or Nemzetközi Repülőversenyek [International Flying Contests], which took place at Rákos-mező, a cavalry exercise field east of Budapest, from 5 to 17 June 1910. Approximately 30 aeroplanes participated, showcasing the growing prowess of European aviation. With a substantial prize pool, this event attracted top aviators from across Europe, including France, Belgium, Germany, Russia, and Austria-Hungary. Over 100,000 spectators attended, witnessing contests in speed, endurance, altitude, and cross-country flights. The event significantly advanced public awareness and support for aviation across Europe.

On 26 February 1910, the Magyar Aero Club was reestablished, soon after an ambitious project was started by the organization—hosting a groundbreaking aviation meet in Budapest. The event received official sanction from the FAI (*Fédération Aéronautique Internationale*) as well as the municipal government. [1]

The *Nemzetközi Repülőversenyek* [International Flying Contests] were held in Budapest, Hungary, at Rákos-mező, a cavalry exercise grounds six kilometers east of Budapest. Approximately 30 machines participated in the competition, which took place from 5 to 17 June 1910. It represented a crucial juncture in the early history of aviation. This seminal event drew substantial crowds and numerous participants, significantly contributing to the acceleration of progress and enthusiasm for aviation throughout Europe and beyond.[2]

Approximately 100,000 spectators attended the event, which effectively demonstrated the wonders of aviation to the public, promoting its potential as a sport, a mode of transportation, and a valuable military resource. The substantial public interest generated by the competition catalyzed further investment and development in the fledgling aviation industry.

The contests fostered international rivalry by bringing together aviators from across Europe, including representatives from France, Belgium, Germany, Russia, and Austria-Hungary. This competitive atmosphere spurred innovation and advancements in aeroplane design, performance, and dependability.

The event offered considerable prize money in 19 distinct categories, encompassing cross-country, endurance, distance, altitude, and speed. The total prize pool of 510,000 kronen, which far surpassed any previous aviation meeting to date. This included a significant contribution of 250,000 kronen from the city of Budapest, it rendered the contest particularly appealing to pilots and designers.[3]

The competition featured esteemed pilots such as Hubert Latham, Louis Paulhan, and Louis Wagner, as well as the noteworthy French aviatrix Raymonde de Laroche. Emerging aeroplane designs, including Antoinette, Farman, Etrich, Voisin, Blériot, and Hanriot machines, were also showcased. Furthermore, domestic aeroplanes and pilots, including Alfred de Pischof, Adolf Warchalowski, Ernő Horváth, Mihály Székely, János Adorján, and Ágoston Kutassy participated in the contest, providing these pioneers with a platform to display their talents and accomplishments.

The Wiener Neustadt Air Meet of 1910

In early June 1910, the Wiener Neustadt Municipal Council proposed to the *Österreichischen Aeronautischen Kommission* [Austrian Aeronautics Commission] to organize an aviation competition called the *I.*

Above: Rákos-mező, Budapest International Flying Contest, 5–17 June 1910. This image features an Antoinette monoplane, which could belong to either Hubert Latham or Eugen Wienczlers, both of whom flew such aeroplanes during the event. Latham's machine, renowned for its elegant design and advanced engineering, was a significant presence at the competition, while Wienczlers piloted an Albatros-Antoinette, which may have undergone an engine change during the meet. The contest marked a major milestone in aviation history, bringing together pioneering aviators from across Europe and showcasing the cutting-edge aviation technology of the time.

Preisfliegen (prize competition) at the Wiener Neustadt airfield over the first three Sundays of July. This contest would exclusively accommodate participants stationed at the Wiener Neustadt aerodrome. Awards would be presented for categories such as speed, altitude, and total flight time, alongside a *Jungfliegerpreis* [junior prize] for the longest flight by a pilot who had not participated in other contests. The Österreichischer Aero-Club would manage the sporting aspects of the event. The proposal was approved, with flight events taking place only on 10, 17, and 24 July.[4] A moderately sized crowd of 50,000 individuals attended the event, a relatively smaller gathering when contrasted with the 300,000 spectators who were present at Louis Blériot's demonstration flight on 23 October 1909. This previous event took place at Simmeringer Heide, situated on the periphery of Vienna.[5]

Prior to the event, the newspapers were invited for a tour of the aerodrome to ensure media coverage. Austrian aeroplane builders including the Autoplan-Werke with their Warchalowski and Pischof designs, as well as Etrich with *Taube* and *Möve* monoplanes were in residence at the field.

Also on view was Baron Constantin von Economo's Voisin and Robert von Lieben's modified Wright-Knoller which would be flown by Vincent Wiesenbach. The Farman, modified by Adolf Warchalowski and now owned by the *k.u.k.Armée,* was present at the aerodrome and would be piloted by *Hptm* Wilhelm Booms, of the Wiener Neustadt aviation field.[6]

Initially, 15 entrants registered for the competition; however, ultimately, six of them did not attempt to fly. Over the final weeks, two more sponsored prizes were introduced, raising the total prize fund to 6,800 kronen.[7] The additional prizes included the *Tüchtigkeitspreis* [proficiency prize] for the pilot who

Above: Rákos-mező, Budapest International Flying Contest, June 5–17, 1910. A Farman-type aeroplane is seen near one of the field's hangars, where notable aviators such as Jorge Chávez, Alfred Frey, Louis Paulhan, Nicolas Kinet, and Adolf Warchalowski competed in various categories. While Warchalowski flew a modified version of the Farman design, known as the Autoplan 'Biplan,' the event prominently highlighted the versatility and influence of the original Farman configuration. These contests not only showcased the advancing technology of aviation but also fostered intense international rivalry. The participation of top pilots and engineers from multiple nations encouraged the exchange of innovative ideas, driving rapid advancements in aeroplane design, performance, and dependability.

made the longest flight on the least favorable day and an unconventional award for the eight-lap flight with the greatest time discrepancy between the first four fast laps and the last four slow laps.[8]

The two air meets had immediate effects on the promotion of aviation. By bringing together a diverse range of aeroplane designs innovations and pilots, the event stimulated further development and innovation in the field. Pilots were able to demonstrate their skills, showcasing the advance of aviation and capturing the public's imagination.

Arguably, the most consequential impact of the Budapest and Wiener Neustadt aviation gatherings, as well as ensuing contests within the Austria-Hungary region, lies in the legacy of pioneering spirit and inspiration. The event's participants played an essential role in shaping the future of aviation in Austria-Hungary, inspiring subsequent generations of pilots, engineers, and enthusiasts to continue pushing the boundaries of flight and develop new technologies.[9]

Endnotes

1. The Magyar Aero Club (Magyar [Hungarian]), originally founded in Budapest in 1902, was a cornerstone of Hungarian aviation and played a vital role in the early development of aeronautics within Austria-Hungary. After its dissolution in 1909, it was reestablished on 26 February 1910, aligning with the burgeoning interest in powered flight. As a member of the *Fédération Aéronautique Internationale* (FAI), the club became an influential entity in promoting aviation.
2. *Vasárnapi Újság* 23. Sz. 1910. (57. Évfolyam.) Budapest, Június 5. [Sunday newspaper №. 23 1910. (Volume 57) Budapest, 5 June], 488-492; *Vasárnapi Újság* 24. Sz. 1910. (57. Évfolyam.) Budapest, Június 12. [Sunday newspaper №. 24 1910. (Volume 57) Budapest, 12 June], 505-509; *Vasárnapi Újság* 25. Sz. 1910. (57. Évfolyam.) Budapest,

Above: Budapest International Flying Contest, June 5-17, 1910, Rákos-mező. A Farman biplane and Hubert Latham's Antoinette monoplane are seen airborne together. At the time, these two aeroplanes represented the state of early aviation technology. Latham's Antoinette, noted for its sleek design and powerful engine, and the Farman biplane, which was a notable and stable aeroplane, earned its reputation as a workhorse during aviation meets.

Június 19. [Sunday newspaper №. 25 1910. (Volume 57) Budapest, 19 June] 529–532.

3. In 1910, 510,000 Austrian-Hungarian kronen would have been worth approximately 102,000 1910-USD. In 1910, 250,000 Austrian-Hungarian kronen would have been worth approximately 50,000 1910-USD.
4. *Wiener Neustadt Nachrichten*, [*Wiener Neustadt News*], 18 July 1909; *Allgemeine Automobil-Zeitung* [General automotive newspaper (Wiener Neustadt Air Meet)] 17 July 1910, 25 September 1910; *Buch des Fluges 1912* [Book of flight 1912] by Hermann Hoernes, 311; *Österreichische Flug-Zeitschrift* [Austrian flight magazine] 1913, 77-78.
5. Keimel, *Österreichs,* 88.
6. *Wiener Neustadt Nachrichten*, [Wiener Neustadt news], 18 July 1909; *Allgemeine Automobil-Zeitung* [General automotive newspaper (Wiener Neustadt Air Meet)] 17 July 1910, 25 September 1910; Österreichischer Aero-Club, 1981. 80 *Jahre Aeroclub*, [Austrian Aero Club, 1981, 80 years of the aeroclub, Vienna] by Franz Pinczolits, 9–12.
7. In 1910, 6,800 Austrian-Hungarian kronen would have been worth approximately 1,360 1910-USD.
8. *Wiener Neustadt Nachrichten*, 18 July 1909.
9. The Österreichische Flugwoche (Austrian Aviation Week), a pivotal event of the 1911 competition season, took place from October 1 to 8 at the Wiener Neustadt airfield. This significant aviation meet attracted widespread attention with a substantial prize pool of 45,200 kroner (approx. $9,040 USD), allocated for achievements in various categories, including the longest endurance flight, the longest distance, the highest altitude, the fastest flights, and cross-country flights to Neunkirchen and Vienna. The competition was preceded by intensive preparations: pilots began training in the last week of September, testing new propellers and tuning engines to enhance speed, along with conducting orientation flights for the cross-country competition. Karl Illner's record-setting flight on 7 October 1911, lasted 4 hours, 4 minutes, and 5 seconds, flying in an Etrich *Taube*. His performance was part of a broader contest that saw Karl Illner win the duration competition, Adolf Warchalowski cover 300 km in an Autoplan biplane for the distance category, *Oblt* Heinrich Bier reach an altitude of 2,018 meters in the *Rennapparat*, and Rudolf Stanger achieve a speed of 85 km/h in an Etrich *Taube*: Zelezny, *Neustadt*, 105–109.

Above: At the Wiener Neustadt airfield in July 1910, attendees witnessed the I. Preisfliegen [First Prize Competition]. Organized to showcase the skill of local aviators and the ingenuity of domestic aeroplane designs, the meet featured categories for speed, altitude, and endurance. Pictured is Karl Illner flying the Etrich II Taube, with the distinctive start number "10" clearly visible on the fuselage. The Taube demonstrated its remarkable capabilities as Illner participated in multiple events, including the endurance competition. The image captures not only the aeroplane in flight but also the captivated audience, reflecting the pride and excitement surrounding the emerging achievements of Austro-Hungarian aviation.

Above: Line-up of 22 aeroplanes, both civilian and military, at the Wiener Neustadt aerodrome before the start of the 18 September 1910 aviation competition. Pilots and owners were formally introduced to Emperor Franz Joseph by *Feldmarschalleutnant* Leopold Schleyer. The diverse range of aeroplanes, featuring designs from Austro-Hungarian as well as international manufacturers, highlighted the technological advancements and competitive spirit of early aviation. This event was a significant milestone in the Austro-Hungarian Empire's adoption and promotion of aviation.

Above: Emperor Franz Joseph, during his inspection at Wiener Neustadt, discusses with Alfred *Ritter* von Pischof the features of his '*Monoplan*', prepared for the aviation competition on 18 September 1910. The emperor's presence and engagement with the aviators highlighted the growing importance of aviation within the empire.

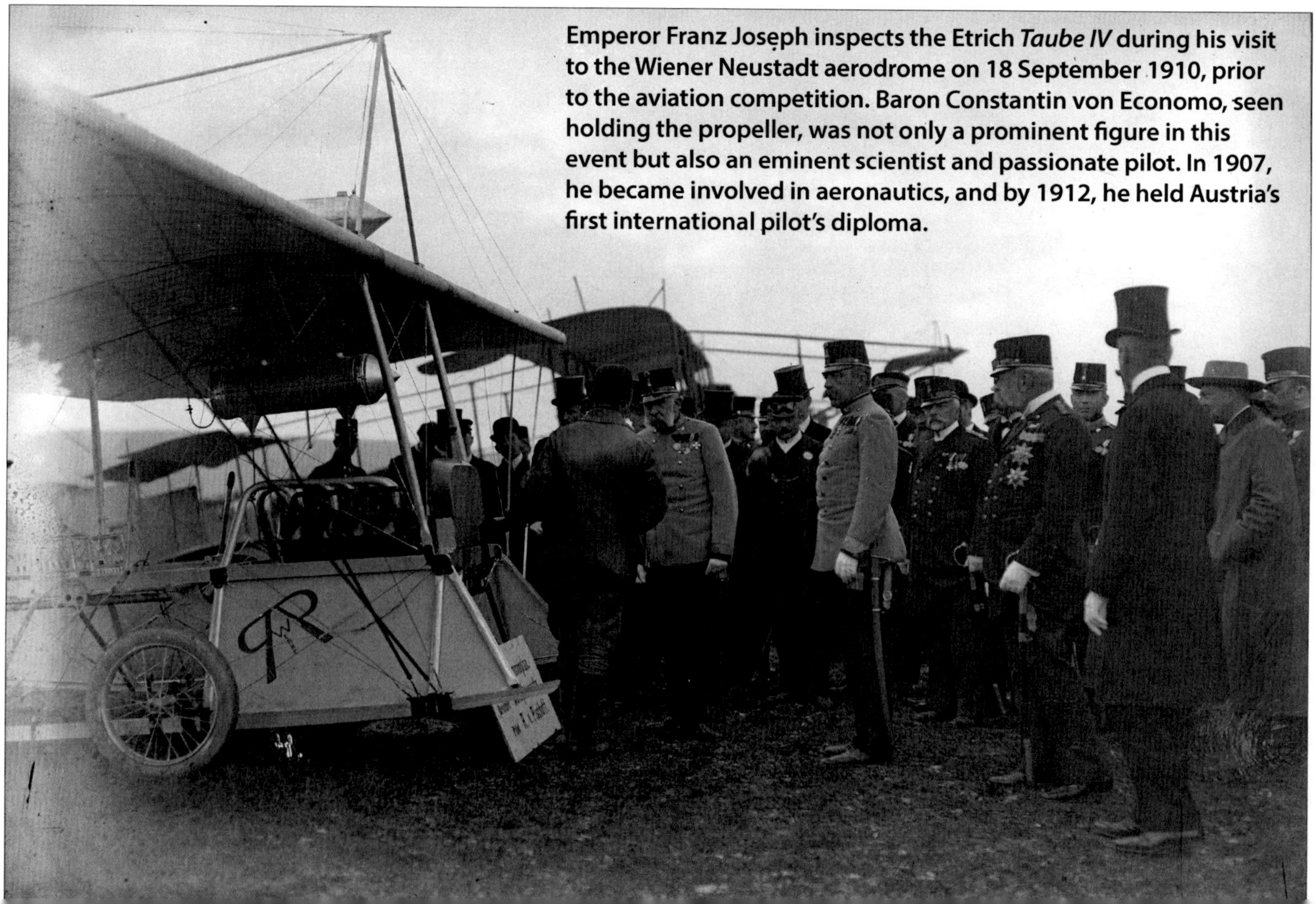

Emperor Franz Joseph inspects the Etrich *Taube IV* during his visit to the Wiener Neustadt aerodrome on 18 September 1910, prior to the aviation competition. Baron Constantin von Economo, seen holding the propeller, was not only a prominent figure in this event but also an eminent scientist and passionate pilot. In 1907, he became involved in aeronautics, and by 1912, he held Austria's first international pilot's diploma.

Above: Emperor Franz Joseph inspects the Warchalowski Autoplan *Biplan* during his visit to the Wiener Neustadt aerodrome prior to the 18 September 1910 aviation competition. In this image, he is seen conversing with Adolf Warchalowski, a pioneer of Austrian aviation whose contributions to early flight technology played a key role in Austria's aviation progress.

Above: The Wiener Neustadt Aerodrome, in 1911, during the peak of its competitive flying season. This colorized postcard captures the *IV Preisfliegen* [Prize Flying Competition] held from 11-18 June, where pilots competed in speed, endurance, and altitude categories. Earlier, on 30 April, a cross-country race to Sopron concluded in a tie between several contenders. The season's grand finale, the Round Flight through Lower Austria from 15-28 August, challenged pilots with a 345 km route divided into five stages. The numbers posted at the grandstands in this image indicate which aeroplanes were currently competing and who was in the air.

Facing Page: This Art Nouveau style promotional poster celebrates the 1911 "*Österreichische Flugwoche*" [Austrian Aviation Week] held at Wiener Neustadt from 1-8 October. The event featured a series of competitions in endurance, speed, altitude, and cross-country flights, attracting members of the royal family. Despite challenging weather conditions, the meet saw significant achievements, including multiple world records in altitude and distance set by pilots like *Oblt* Heinrich Bier and Adolf Warchalowski. With total prizes worth 45,200 kronen, the event marked a pivotal moment in Austria's aviation history, showcasing technological advancements and the pilots' daring in the face of adverse elements.

FRITZ-SCHONPFLUG
KUNSTDRUCK
A. LUIGARD

7. Building the Air Service

Above: In the cockpit of the "Etrich VIII" stands *Oblt* Miecislaus Miller, the first military pilot of Austria-Hungary. Standing in front of the aeroplane, from left to right: *Oblt* Philipp Blaschke *Ritter* von Zwornikkirchen, *Oblt* Karl Stohanzl, Eduard Fischer (Director of ÖDMG), *Rittm* Ludwig Schmidl, Camillo Castiglioni (Director of MLG), *Hptm* István Petróczy von Petrócz, and Karl Illner. This MLG-built *Taube* would become the military's first ordered aeroplane. The photograph was taken at Wiener Neustadt on 2 April 1911. *Credit: Bohumír Kudlička*

In light of the burgeoning aviation industry and significant media coverage, the Vienna War Ministry could not remain idle. The national press particularly highlighted Etrich's *Taube* aeroplane. Consequently, on 19 April 1911, the same day the first aeroplane purchased by the Austrian military, an Etrich *Taube* (*Etrich VIII*), officially entered service—marking the commencement of dedicated military pilot training in Austria-Hungary—the War Ministry established the first *k.u.k.* military pilot school in Wiener Neustadt, named *Flugmaschinen Instruktionsabteilung* [Flying Machine Instruction Detachment], commanded by *Hptm* István Petróczy von Petrócz. In May 1911, 27 selected volunteer officers enrolled in the inaugural military pilot training course in Wiener Neustadt.[1]

The potential utility of aeroplanes in military situation assessments remained to be demonstrated. However, two significant exercises in 1911 provided convincing evidence of the importance of airborne reconnaissance.

In the cavalry, the exercise was carried out in southern Hungary from 21–23 August 1911 and the exercise of the *V. Korpskommando* [*k.u.k* V. Corps] in Hungary from 3–6 September 1911. Both the blue and red teams each possessed a *Militärischer Flugmaschinen Park* [Military Flying Machine Park] and a civilian counterpart. The Civilian Flying Machine Parks, as the name suggests, were staffed by civilian pilots flying civilian aeroplanes, but they were overseen by a military officer. This arrangement arose due to a shortage of military pilots, necessitating civilian involvement under military supervision. *Hptm* Petróczy commanded the Military Flying Machine Parks in both exercises.

The military pilots operated *Taube* aeroplanes (initially three, later four), while the civilians used both *Taube* planes and their privately owned aeroplanes

Above: Members of the extensive ground crew of the *k.u.k. Luftschifferabteilung* are seen working with the *Taube* aeroplane named '*AFRA*.' At this stage, Austro-Hungarian military aeroplanes were identified by unique names stenciled prominently on their fuselages—a practice that would later transition to numerical designations as the fleet expanded. The name '*AFRA*' is also marked on the transport cart, underscoring the importance of precise logistical coordination in maintaining and supporting aeroplane operations. The *Taube*, an early reconnaissance and training aeroplane, required dedicated maintenance and preparation, often involving large teams like those seen here. Extensive field operations included tasks such as rigging repairs, engine checks, and fuel replenishment, all critical for ensuring readiness in austere and challenging conditions. Assigned to '*AFRA*' was *Oblt* Karl Stohanzl, a *Feldpilot* responsible for flying the aeroplane during its service. *Credit: Koloman Mayerhofer*

(three in total). Both the military and civilian aviators delivered exceptional reconnaissance results. *Chef des Generalstabs* Conrad expressed his gratitude to the aviators and acknowledged the extraordinary value of aviation for higher commands.[2] Following the exercises, the participating military pilots received the title of *k.u.k. Feldpilot* [Field Pilot].[3]

Two events in 1911 further catalyzed the development of the *k.u.k.* Air Service. First, Italy's war against Turkey in Tripolitania (present-day Libya) showcased the potential of aviation in warfare. Second, War Minister *General der Infanterie (GdI)* [General of Infantry] Baron von Schönaich, who opposed aviation, was replaced by *GdI* Moritz *Freiherr* Auffenberg von Komarów, a proponent of aeroplanes.[4]

In Libya, the Italians pioneered the use of aeroplanes in warfare by employing an Etrich *Taube* and the Blériot XI for aerial reconnaissance and conducting opportunistic bombing.[5] While the aerial reconnaissance was successful, the bombing was not particularly effective as only small handheld ordnance was employed.

As one of his initial actions, the new War Minister initiated a significant reorganization of the *k.u.k. Luftschifferabteilung*, Fischamend, on 1 October 1911, to better align it with the needs of a future *k.u.k.* Air Service. Two departments within the Aeronautics Department were created for aviation: the *Unterricht Abteilung* [Instruction Department] and the *Technische*

Above: The *k.u.k. Luftschifferabteilung Taube* aeroplane, named *'ARAM,'* piloted by *Rittmeister* Hans *Ritter* Umlauff von Frankwell, is seen here with an observer at the Wiener Neustadt military aerodrome. The *Taube's* distinctive bird-like design was inspired by the Alsomitra macrocarpa (commonly referred to as the Zanonia seed), whose gliding stability was first studied in the 1890s by Friedrich Ahlborn and later influenced aviation pioneers like Igo Etrich. As a domestically designed and constructed machine, its aerodynamic features, including wing warping for lateral control and a lightweight structure, underscored its versatility as an early reconnaissance and training aeroplane. A ground crew mechanic is seen attending to the engine, while the observer remains in his officer's cap—an interesting detail that suggests pre-flight preparations. In this period of early military aviation, it was not uncommon for officers to wear their standard service caps. during ground activities. Before takeoff, these caps would typically be stowed in a secure location due to practicality and safety considerations in the open cockpit.

Abteilung [Technical Department]. The existing *Luftschiffer* detachment remained, but it received minimal attention. The chain of command was not altered; the Aeronautics Department continued to operate under the Transport Troops Brigade. However, the fortress balloon service, which had previously been part of the Aeronautics Department, was separated and reassigned to the fortress artillery.[6]

On 11 December 1911, *Chef des Generalstabs* Conrad once again appealed to the new war minister to establish a dedicated military aeroplane service, comprising peacetime *Flugpark*s [airparks] and wartime field units. Conrad proposed the creation of 40 field units with a total of 200 aeroplanes, complete with air crews, maintenance, ground equipment, and staff. Furthermore, he suggested that a list of stringent technical requirements for military aeroplanes be submitted to the industry.[7]

Despite War Minister von Auffenberg's support, he was unable to secure political approval for the necessary funding. Conrad, disheartened by this outcome, sought the emperor's assistance in two separate instances, but to no avail. The *k.u.k.* military administration was constrained by the constitutional rules of the multi-ethnic Austro-Hungarian Monarchy, which was characterized by ongoing disputes over finances and national interests between the western Austria-dominated Cisleithanian[8] half and the Hungary-dominated Transleithanian[9] half of the empire. As a result, decision making was slow, susceptible to political interference, and often unsatisfactory.[10]

Above: The commander, *Hptm* István Petróczy von Petrócz, together with the four training officers and the eleven trainee pilots, are grouped in front of the Etrich *Taube 'ADLER'* on the airfield at Wiener Neustadt during the first flight course of the Austro-Hungarian Army's "*Flugmaschinen-Instruktionsabteilung*," held from 20 April to 16 September 1911. In the background, the Emperor's Pavilion is visible on the left, with a line of hangar buildings on the right. The officers, from left to right: *Lt* Anton Venczel, *Oblt* Karl Stohanzl, *Lt* Rudolf Holeka, *Lt* Erich Oelwein, *Hptm* István Petróczy von Petrócz (commander of the station, 5th from right), *Oblt* Eduard Babouczek, *Oblt* Robert Eyb, *Rittm* Ludwig Schmiedl, *Oblt* Miecislaus Miller, *Oblt* Eduard Nittner, *Lt* Johann Riedlinger von Kastrenberg, *Oblt* Philipp Blaschke *Ritter* von Zwornikkirchen, *Lt* Viktor Schünzel, *Oblt* Kurt Wilhelm von Helmfeld, *Lt* Karl von Banfield, and *Fregattenleutnant* Heinrich Huss. (Photographer: Josef Ferber, Wiener Neustadt). *Credit: Koloman Mayerhofer*

Establishing a national aviation industry was costly, and without significant government procurement, neither entrepreneurs nor investors were likely to participate. By the end of 1911, the Vienna War Ministry had purchased a mere ten aeroplanes in total: eight Etrich *Taube*, one Warchalowski biplane, and one of Pischof's Autoplan *Monoplan* machines.[11]

The reorganization of the Aeronautics Department spanned from 1 October 1911 until the early weeks of 1912. On 24 April 1912, the unit commander, *Hptm* István Petróczy von Petrócz was replaced by *Obstlt* Milan (Emil) Uzelac, a highly skilled and assertive officer from the *Genie Stab* [Engineer Staff].[12] The appointment of *Obstlt* Uzelac to lead the Aeronautics Department proved to be a fortunate decision. Born on 26 August 1867, in Komárom, Hungary, Uzelac was orphaned at an early age and raised in an Offizierswaiseninstitut [Officers Orphan Institute], which offered education and upbringing to the orphaned children of Austro-Hungarian military officers. Uzelac displayed exceptional talent and diligence, leading to his selection for an officer career. After studying at the Technical Military Academy in Vienna, he demonstrated his capabilities in subsequent technical assignments.

Immediately after assuming command of the restructured Aeronautics Department, Uzelac, who was 45 at the time, focused on learning to fly. Reportedly, he would eventually personally pilot every type of aeroplane in the *k.u.k.* service. His officers and men admired him, affectionately referring to him as *Unser*

Above: *Obstlt* Milan (Emil) Uzelac, the newly appointed commander of the *k.u.k. Luftschifferabteilung*, seen here in the cockpit of a military aeroplane. Uzelac, a skilled and highly respected officer from the *Genie Stab* [Engineer Staff], took command on 24 April 1912 and immediately began flying training. His dedication to aviation made him one of the key figures in the development of the *k.u.k. Luftfahrtruppen*.

Uz (our Uz). Ultimately, he became the true creator of the *k.u.k. Luftfahrtruppen*.[13] *Obstlt* Uzelac remained in command until the end of World War I in 1918. On 1 November 1914, he was promoted to *Oberst (Obst)* [Colonel], and in June 1918, he was appointed *Generalmajor*.[14]

The outbreak of the First Balkan War in the fall of 1912 offered another opportunity to advance the *k.u.k. Luftschifferabteilung. Chef des Generalstabs* Conrad once again petitioned the War Minister to substantially develop the scattered and inadequately sized Aeronautics Department. This time, his efforts were successful.

The 1912 Balkan War, involving the anti-Turkish alliance of Bulgaria, Serbia, Greece, and Montenegro against a weakened Turkey, occurred on Vienna's periphery and directly affected Austro-Hungarian political interests in the Balkans. With the threat of being drawn into the conflict, Vienna hastily allocated funds for the enhancement and modernization of its army. Most services, including the Aeronautics Department, benefited from this allocation. Under the pressure of the Balkan crisis, the War Ministry finally granted the augmentation for aviation. In this context, Das Offiziersfliegen [Flight Officers Meeting], held on 6 October 1912, served as a crucial public and military showcase of these enhancements. This event not only demonstrated the advanced capabilities of the Aeronautics Department but also played a key role in affirming the strategic importance of the air service in modern warfare.[15]

For the first time, a significant number of aeroplanes were ordered, airfields and ground infrastructure were constructed, and logistical facilities were established. A considerable sum of 7.9 million kronen was allocated for aviation, a substantial amount for Austro-Hungary, though modest compared to Germany and Russia. Nevertheless, this marked a significant stride forward for *k.u.k.* aviation. For the following year, 1913, another 4.6 million kronen were scheduled for expenditure,

Above: *Feldpilot Oblt* Philipp Blaschke *Ritter* von Zwornikkirchen and passenger *Oblt* Karl Banfield, himself a certified pilot (№ 27) since 23 September 1911, are pictured after setting a world altitude record of 4,360 meters on 29 June 1912 at the Aspern airfield in Vienna. Flying what would later be named the Lohner *Pfeilflieger* "*Aspern*," equipped with a 120 hp Austro-Daimler engine, Blaschke and Banfield achieved this remarkable feat during the *Erste Österreichische Flugwoche* [First Austrian Aviation Week]. Their flight also set a world record for rate of climb with a passenger, reaching 1,000 meters in just 6 minutes and 15 seconds. The record-breaking flight lasted approximately 1 hour and 36 minutes. Battling thin air and freezing temperatures that dropped to -10°C at altitude, the duo faced significant challenges. At one point, Banfield had to temporarily take over the controls while Blaschke massaged his hands to restore circulation. Upon landing, their accomplishment was met with great fanfare. The aeroplane was adorned with a laurel wreath, and Archduke Josef Ferdinand personally congratulated them.

indicating that politicians and senior military officials responsible for the budget recognized the importance of aviation for future warfare.[16]

During the 1912 fall maneuvers, aviators further demonstrated their value in informing the decision-making of higher army commanders. Air units had definitively transcended the experimental stage and earned considerable recognition as a valuable and essential reconnaissance service.[17]

In acknowledgment of their importance, Emperor Franz Joseph approved a distinguished badge for aviators on 11 September 1912. The badge featured an oak-leaf-wreathed raptor bird with outstretched wings beneath the *k.u.k.* imperial crown, similar to the German flyer badge. However, the War Ministry, in its typical delay, did not issue regulations until 30 May 1913, specifying that the badge was to be worn exclusively by qualified *Feldpiloten*.[18]

In 1912, the *k.u.k. Luftschifferabteilung* experienced its most significant expansion to date. The War Ministry awarded contracts for the purchase of 50 aeroplanes, which included 18 Etrich *Taube* aeroplanes (with four *E* double-control versions) and 32 Lohner *Pfeilflieger* (some of which were to be delivered by April 1913). The cost of one *Taube* airframe, excluding the engine, was approximately 5,500 kronen, whereas a *Pfeilflieger*, also without an engine, was priced at 17,500 kronen.[19]

Above: Radul Milkov in an Albatross F2 aeroplane, moments before the first combat flight in Europe on 16 October 1912, during the First Balkan War. Milkov, along with fellow Bulgarian pilot Prodan Tarakchiev, conducted a 50-minute reconnaissance flight over the besieged Edirne fortress. This groundbreaking flight, carried out just days after the war began, left the Ottoman military astonished and marked a pivotal moment in military aviation history.

Obstlt Uzelac focused on pilot training but also emphasized the training of airship crews and captive balloon aeronauts. In addition to military activities, civilian associations promoting aviation played a crucial role in the early development of Austro-Hungarian aviation. For the patriotic elite in Vienna, civilian and military aviation endeavors were considered inseparable.[20]

An influential group of wealthy and high-ranking individuals, including members of the Habsburg royal family, established the Österreich-Ungarisches Luftflotten Komitee [Austro-Hungarian Air Fleet Committee] to raise funds for the creation of an air service, similar to initiatives in Germany and Russia. This effort was strongly supported by the officers of the Aeronautics Department. By 1914, the committee had raised 1.4 million kronen and presented it to the War Ministry, although the fate of these funds remains unclear in the fog of the emerging war.[21]

Numerous aero clubs were formed in the prewar decade, with the Österreichischer Aero-Club being particularly influential due to its political connections. Many members, including industrialists and general officers, had access to top political or military decision makers. The club successfully campaigned for the establishment of a municipal airfield in Wien-Aspern, a suburb of Vienna. Austro-Hungarian civilian pilots were aware that they would don military uniforms and serve in *k.u.k.* field units in the event of military call-up.[22]

During the pivotal transition period of 1912–1913, notable shifts occurred within the *k.u.k.* military hierarchy, setting the stage for advancements in aviation. *Feldmarschalleutnant (Fmlt)* [Lieutenant Field Marshal] Leopold *Freiherr* Schleyer von Pontemalghera, who had been recently elevated in rank and ennobled, was succeeded as commander of the Troop Transport Brigade by *Genmjr* Wenzel Tertain. This change coincided with a broader reshuffle at the highest levels, with *Feldzeugmeister* (*FZM*) [General of the Artillery] Alexander *Freiherr* von Krobatin replacing War Minister von Auffenberg.

Around the same time, a young, low-ranking *Generalstab* officer, *Lt* Alexander Löhr, was assigned

Above: Ottoman pilots in a French-built REP aeroplane, designed by the aviation pioneer Robert Esnault-Pelterie and introduced prior to the Balkan Wars (1912–1913). The REP aeroplanes, featuring a steel-tube fuselage and monoplane configuration, exemplified modern aviation design of the era. Acquired as part of the Ottoman Empire's initial steps toward establishing a military aviation force, these aeroplanes were employed during the Balkan Wars for reconnaissance missions. Their use represents one of the earliest documented instances of Ottoman military aviation, highlighting the Empire's early interest in integrating aeroplanes into its strategic capabilities.

to the Troop Transport Brigade, making a significant contribution to aviation. He developed the mobilization plan for the Aeronautics Department, which was to take effect in August 1914.[23] Notably, Löhr had a distinguished career in the Aeronautics Department.[24] The growing demand for pilots led the War Ministry to plan for four additional flying schools in addition to Wiener Neustadt. In 1913, the following locations were planned for pilot schools: Görz, Fischamend, Wiener Neustadt, Wien-Aspern, and Ujvidék, Hungary.[25] However, the plan's implementation was slow, and by August 1914 there was only one flying school in operation in Austria: Wiener Neustadt. Wien-Aspern would not be operational as a school until November 1914, and Fischamend would follow in January 1915.[26]

The Second Balkan War in June 1913 intensified Vienna's threat perception, as the Balkans had always been a politically volatile region. The hostile posture of Serbia, which was allied with Montenegro and backed by Russia, increasingly threatened the *k.u.k.* Monarchy. Many influential political and military decision makers in Vienna concluded that war with Serbia was inevitable.

Chef des Generalstabs Conrad consistently advocated for the technical modernization and expansion of the *k.u.k. Armée*, including the augmentation of the Air Service, in preparation for a war against Serbia. By the end of 1913, he once again presented his grand augmentation plan for the Air Service to the war minister. His suggestions are documented in his memorandum of January 1914.

The plan aimed to increase the Aeronautics Department to a total of 240 aeroplanes by 1916, grouped into 40 mobile aviation units of six aeroplanes each, complete with all necessary ground equipment and personnel. This plan required purchasing aeroplanes and ground equipment, as well as training sufficient pilots and technical support personnel. In peacetime, *Flugparks* were to be established as necessary cadre units, housing all aeroplanes, ground equipment, and other resources for their planned number of field units to be mobilized, known as *Fliegerkompanien* or *FliK*s [Flyer Companies].[27]

Conrad reported a total of 84 field serviceable aeroplanes, along with 35 training aeroplanes.28 However, this assessment seemed to have euphemistically glossed over the true status of operational readiness. In reality, the *Flugparks* had only 48 operationally ready aeroplanes, 33 of which were Lohner *Pfeilflieger* that were subsequently grounded due to several fatal crashes. Investigations revealed that the *Pfeilflieger* wings were prone to breaking because of structural weaknesses. The Lohner company had reduced the required breaking loads, contrary to the War Ministry's requirements, leading to fatal consequences.

Meanwhile, *Chef des Generalstabs* Conrad no longer appeared to encounter disapproval from the Minister of War. However, his plan for army reinforcement could not be fully implemented. Austria-Hungary had

I. Oesterreichisches Flugfeld in Wiener-Neustadt.

Uebungsordnung

für das

zu Ehren des Allerhöchsten Namensfestes

am 6. Oktober 1912, nachmittags, in Wr. Neustadt

stattfindende

Offiziersfliegen.

1. Abflug dreier Flugzeuge nach Sopron (Ödenburg). Dem ersten Flugzeug gibt noch ein viertes eine Strecke hindurch das Geleite.
   Abflug 3 Uhr — Min.: „Asra", Führer: Oberleutnant Eyb.
   „Aar", „ Oberstleutnant Ucelac (gibt das Geleite).
   3 Uhr 5 Min.: „Annita" „ Oberleutnant Holeka.
   3 Uhr 10 Min.: „Adjutant", „ Leutnant Flassig.
   (Auf dem Kavallerie-Exerzierplatz in Sopron Zwischenlandung).
2. Ankunft zweier Flugzeuge aus Fischamend. Abflug in Fischamend 2 Uhr 30 Min. nachmittags.
   „Bravo", Führer: Oberleutnant v. Blaschke.
   „Amourette", „ Leutnant Veczel.
3. Schulfliegen.
   a) Vorführung einer Pilotenprüfung.
   „Afra", Führer: Oberleutnant Stohanzel.
   b) Gleitflug, Luftlandung, dann Landung mit motorischer Kraft.
   „Ajax", Führer: Oberleutnant v. Wilhelm.
   c) Gleitflug, Sturzflug und Landung mit ganz abgestelltem Motor.
   „Amazone", Führer: Leutnant Schünzel.
4. Höhenflug auf 1000 m über dem Erdboden. Die erreichten Höhenstufen von 100 m, 200 m, 300 m etc. werden durch Abschießen von Leuchtkörpern bemerkbar gemacht.
   „Arám", Führer: Rittmeister v. Umlauff.
5. Besondere Flugübungen.
   a) Kurven- und Achter-Fliegen.
   „Achilles", Führer: Oberleutnant Nittner.
   b) Spiralen im Gleitflug.
   „Allons", Führer: Oberleutnant Miller.
   c) Schleifen im Gleitflug.
   „Bravo", Führer: Oberleutnant von Blaschke.
6. Fliegen im Geschwader: 10 Flugzeuge unter Führung des „Aar", Führer: Oberstleutnant Ucelac.

Alle Flüge werden mit Passagiere durchgeführt. **Das Fliegen findet bei jeder Witterung statt.**

Preis 20 Heller.

Das Reinerträgnis fließt dem Fond zur Schaffung der österr. Luftflotte zu.

Above: Original German program sheet titled '*Das Offiziersfliegen*,' outlining the "Exercise Program for the Officers' Flight Meeting" held on October 6, 1912, in Wiener Neustadt. The document details the scheduled flights and aerial maneuvers conducted to honor the Name Day of Emperor Franz Joseph I, a significant event in the Austro-Hungarian military calendar. It highlights the early development of organized military aviation in the Austro-Hungarian Empire, underscoring the growing role of aeroplanes in both training exercises and ceremonial functions during this formative period.

Austrian Airfield in Wiener-Neustadt

Exercise Program

Held in Honor of the Highest Name Day

on 6 October 1912, in the Afternoon, in Wiener-Neustadt

Officers' Flight Meeting

1. Departure of three aircraft to Sopron (Ödenburg). The first aircraft will be accompanied by a fourth one for part of the way.
   Departure 3 hr. 0 min. - ***"Asra"***, Pilot: Oberleutnant Eyb.
   "Aar", Pilot: Oberstleutnant Ucelac (escort).
   3 hr. 5 min. - ***"Annita"***, Pilot: Oberleutnant Holeka.
   3 hr. 10 min. -***"Adjutant"***, Pilot: Leutnant Flassig.
   (Intermediate landing at the Cavalry Exercise Field in Sopron).

2. Arrival of two aircraft from Fischamend. Departure from Fischamend at 2 hr. 30 min.
   "Bravo", Pilot: Oberleutnant von Blaschke.
   "Amourette", Pilot: Leutnant Veczel.

3. Training Flights.
   a) Demonstration of a pilot examination.
   "Afra", Pilot: Oberleutnant Stohanzel.
   b) Glide flight, landing, then landing with engine power.
   "Ajax", Pilot: Oberleutnant von Wilhelm.
   c) Glide flight, dive, and landing with the engine fully off.
   "Amazone", Pilot: Leutnant Schünzel.

4. Altitude flight to 1000 meters above the ground.
   The heights reached at 100 m, 200 m, 300 m, etc., will be indicated by the firing of signal flares.
   "Aram", Pilot: Rittmeister von Umlauft.

5. Special Flight Exercises.
   a) Curve and figure-eight flying.
   "Achilles", Pilot: Oberleutnant Nittner.
   b) Spirals in a glide flight.
   "Allons", Pilot: Oberleutnant Miller.
   c) Loops in a glide flight.
   "Bravo", Pilot: Oberleutnant von Blaschke.

6. Squadron Flying: 10 aircraft under the leadership of *"Aar"*, Pilot: Oberstleutnant Ucelac.

All flights will be conducted with passengers. The flying will take place in any weather.

Price: 20 Heller.

The net proceeds will go to the fund for the creation of the Austrian Air Fleet.

Above: Facsimile translation of the "Exercise Program for the Officers' Flight Meeting," outlining the series of aerial demonstrations and training flights held in Wiener Neustadt on 6 October 1912. The document provides a detailed overview of the event, featuring named aeroplanes, their pilots, altitude tests, maneuver exercises, and squadron formations. These activities exemplify the Austro-Hungarian military's focus on precision, discipline, and the integration of aviation into formal military operations. Proceeds from the event were allocated to the development of the Air Fleet, underscoring its dual role in advancing aeronautical training and cultivating public support for military aviation.

Above: On 6 October 1912, Wiener Neustadt hosted *Das Offiziersfliegen*, a noteworthy assembly of Austro-Hungarian military aviators participated. Organized by the *k.u.k. Luftschifferabteilung*, the event blended spectacle with purpose, highlighting piloting skills and precision maneuvers. The *Taube* aeroplanes, each crewed by a pilot and an observer, are arranged on the field from right to left: *ASRA* (*Oblt* Robert Eyb), *AMAZONE* (*Lt* Viktor Schünzel), *ARÀM* (*Rittm* Hans *Ritter* Umlauff von Frankwell), and *AJAX* (*Oblt* Kurt von Wilhelm). The meeting demonstrated the advanced capabilities of Austro-Hungarian military aviation and affirmed its growing importance within the armed forces. *Credit: Christian Reiterer*

Left: The *Militär-Fliegerabzeichen* [Military Pilot's Badge], awarded to *k.u.k. Luftfahrtruppen Feldpiloten* [Field Pilots], reflects a blend of imperial heraldry and the emerging identity of military aviation in the Austro-Hungarian Empire. Approved by Kaiser Franz Joseph on 11 September 1912, the badge features an eagle with outstretched wings—an enduring symbol of power, vigilance, and mastery of the skies—encircled by an oak-leaf wreath, representing strength and endurance. Above the eagle rests the imperial crown, invoking loyalty to the Habsburg monarchy, while a shield at the base bears the initials "FJI" for Franz Joseph Imperator, grounding the design in dynastic authority. Though regulations for wearing the badge were not issued until 30 May 1913, its eventual adoption marked it as a distinguished emblem of achievement. Reserved for qualified *Feldpiloten*, the badge encapsulates the duality of tradition and modernity. *Credit: Robert Pandis*

Flugpark (Flight Park)	Personnel	*Flugzeuge* (Aeroplanes)
Flugpark 1 - Fischamend	*Kommandant*: *Obltn* Eyb, *Feldpilot* (FP) *Hptm* Dworzak, *Technischer Offizier* (TO) *Lt* Cizek	Atair, Amazone, Amourette, Anny
Flugpark 2 - Görz	*Kommandant*: *Obltn* Stohanzl, FP Obltn Holeka, FP *Obltn* Kenese, TO *Lt* Hauer, Schüler: *Lt* Perini, *Obltn* von Nemec, *Lt* Schönpflug, *Lt* Schindler	Adler, Falke, Habicht, Condor
Flugpark 3 - Ujvidek	*Kommandant*: *Hptm* von Petrozy, FP *Obltn* von Wilhelm, *Obltn* Banfield, TO *Obltn* Flassig	Afra, Ajax, Aram, Anita
Flugpark 4 - Krakau	*Kommandant*: *Obltn* Oelwein, FP *Lt* von Riedlinger	Aar, Bravo
Flugpark 5 - Przemysl	*Kommandant*: *Obltn* Miller, FP *Lt* Venczel, TO *Obltn* Baar	Allons, Asra, Armin, Adjutant
Flugpark 6 - Sarajevo	*Kommandant*: *Obltn* Nittner, FP *Lt* Schünzel, TO *Obltn* Javor	Bora, Butterfly
Flugmaschinenersatzpark (Aeroplane Reserve Park) - Wiener Neustadt	*Kommandant*: *Mjr* von Umlauft, FP *Obltn* von Blaschke, TO *Lt* Libowitzky, Schüler: (students): *Hptm* Andrich, *Obltn* Stiasny, *Obltn* Libano, *Lt* Schar, *Obltn* Elsner, *Lt* Rzemenowsky, (etc.)	F1, F2, F3, F4

Above: As part of a strategic reorganization in December 1912, the Austro-Hungarian military established *Flugparks* [Flight Parks] across key locations. Each *Flugpark* was staffed with commanders, *Feldpiloten*, and *Technische Offiziere* [Technical Officers], with various aeroplanes such as the Etrich *Taube* variants allocated accordingly. This arrangement ensured the readiness and maintenance of machines, laying the foundation for *k.u.k. Luftfahrtruppen* operations in the pre-war period.

Above: Lineup of military Taube aeroplanes at Wiener Neustadt airfield, circa 1913. Following the relocation of civil aviation activities to Vienna/Aspern, the military leased the airfield at a favorable rate, transforming it into a key training and operational site. Two aviation units, known as '*Flugparks*,' were stationed here, and military pilots received instruction at the *Flugmaschinen Instruktionsabteilung* [Flying Machine Instruction Department]. As the First World War loomed, the airfield's strategic importance grew, leading to significant expansion, including new hangar facilities and the establishment of the *Fliegeroffiziersschule* [Flying Officers' School] for the training of observers and technical officers. Wiener Neustadt also became home to the Oesterreichische Flugzeugfabrik AG (Oeffag), founded in 1915, which produced over 500 license-built Albatros D.III fighters, as well as aero-engines developed by the Oesterreichische Daimler Motoren Gesellschaft under the direction of Ferdinand Porsche.

sufficient armament industries for its ground forces but lacked a developed aeroplane and engine industry. Despite the national and international successes of its aviators, the absence of substantial orders for aeroplanes meant that no notable *k.u.k.* aviation industry had formed. Some companies demonstrated their capacity to produce advanced aeroplanes and engines, but they had not received orders for significant deliveries from the War Ministry.

The War Ministry's policy was characterized by

Flugparks

Flugpark Number	Location	Qualified Military Aeroplanes	Training Aeroplanes
1	Fischamend	9 (3 Etrich Taube, 1 Lohner B, 5 Lohner D)	9 (1 Lohner, 1 Lohner B, 7 Etrich Taube)
2	Wiener Neustadt	9 (2 Etrich Taube, 7 Lohner B)	7 (7 Etrich Taube)
13	Wiener Neustadt	6 (6 Etrich Taube Version 1912)	n/a
3	Goerz	3 (1 Lohner, 1 Lohner C, 1 Lohner E)	4 (4 Etrich Taube)
4	Mostar	1 (1 Lohner B)	n/a
6	Mostar (Training)	n/a	n/a
5	Ujvidék	4 (4 Etrich Taube 1912)	7 (7 Etrich Taube 1911)
7	Krakau	4 (4 Lohner B)	2 (2 Etrich Taube 1911)
8	Aspern (Vienna)	6 (6 Lohner B)	n/a
14	Aspern (Vienna)	6 (6 Lohner B)	(2 Etrich Taube 1911)

Above: The Emperor's Pavilion at Wiener Neustadt, originally built in June 1910 as a gift from Archduke Friedrich. This two-story structure became a central feature of the airfield, symbolizing imperial support for aviation. After Emperor Franz Joseph I's visit on 18 September 1910, the pavilion was renamed the "Emperor's Pavilion" in his honor. The building became a prominent site for military aviation events and an emblem of the growing significance of the airfield to the Austro-Hungarian military.

indecisiveness due to the delicate political balance among the conflicting interests of the major groups within the multi-ethnic Austria-Hungary. High-ranking decision-makers were accustomed to taking minimal action, maintaining aeroplane production capabilities without offering growth prospects. Emblematic of this dysfunctional policy for the development of a substantial aeroplane industry, the Aeronautics Department operated a limited production facility in Fischamend in 1913, known as the *Flugzeugfabrik des Fliegerarsenals* [Aeroplane Factory of the Aviation Arsenal]. This facility was not capable of mass production, functioning instead as a workshop for repair and occasional production of individual aeroplanes.

In July 1914, as Austro-Hungary prepared for war, only the Lohner factory and Ungarische Flugmaschinen A.G. were capable of accepting orders for airframes, while Österreichische Daimler Motoren A.G. and Österreichische Industriewerke received orders for engine production. The national production base could not meet the required volumes, necessitating assistance from Germany, whose aeroplane and engine industry was more extensive, primarily due to stronger military support for production and procurement policies, compared to that of Austria-Hungary[29]

A comparison with the German ally highlights the disastrous lack of Austro-Hungarian aeroplane production in preparation for war. Between 1912 and 1914, Germany had promoted the development of a robust aeroplane and aviation engine industry, with over 700 orders. Consequently, there were three principal factories—Albatros, Aviatik, L.V.G.—and a number of secondary ones, along with engine manufacturers Daimler, Argus, Benz, Oberursel, and others. At the beginning of the war, Germany's production capabilities

Above: Colorized photomontage postcard of Wiener Neustadt, capturing the scene at the airfield in June 1911. The image looks towards the wind gauge tower, designed by Prof. Ludwig Schmidt, adjacent to the Emperor's Pavilion on the right. Several hangars are visible to the left, part of the airfield's extensive infrastructure. In the sky, an Etrich *Taube* and a *Renntaube* [Racing Dove] can be seen in flight over the aerodrome. This postcard offers a glimpse into the bustling activity at Austria's first airfield, which by this time had 33 hangars and hosted both civilian and military aviation activities.

Left: *Oblt* Alexander Löhr, a *Generalstab* officer, made crucial contributions to Austro-Hungarian military aviation. Löhr was responsible for developing the mobilization plan for the *Luftschifferabteilung*, which played a significant role in the lead-up to World War I.

Above: Bulgarian air crew prepare for a mission to drop bombs by hand on Adrianople from their Bleriot XI aeroplane, during the Balkan War, circa 1913.

were estimated to be around 100 airframes and 75 engines per month.[30]

Financial figures for the major European powers corroborate these facts. From 1909 to 1913, European powers spent the following amounts on air armament (in USD million): [31]

Germany:	28
France:	22
Russia:	12
Italy:	8
Austria:	5
Great Britain:	3
Bulgaria:	0.6
Romania:	0.2

Although the figures are only estimates, they provide a clear illustration of the Vienna War Ministry's relatively low expenditure on air armament compared to other major European continental powers. An examination of the expenditures in relation to the respective population totals reveals an uncomfortable reality for Vienna: Austria-Hungary had approximately 40 million inhabitants, while Germany had around 60 million. Given the 3-to-2 ratio of the German to the Austro-Hungarian population, Vienna should have spent 16 million USD (equivalents) on the air arm. However, they fell short of this figure by a factor of three.

Obstlt Uzelac, commander of the Aeronautics Department, was the ideal officer for *Chef des Generalstabs* Conrad to vigorously execute expansion plans. However, with limited resources at his disposal, he could only make the most of what little he had. Despite the insufficient quantity of aeroplanes and technical equipment, Uzelac pressed on with aviator training.

On 21 April 1913, detailed regulations were issued outlining the qualifications and skills required for pilot training candidates. In addition to practical flying skills, candidates were required to demonstrate knowledge in aviation related meteorology, motoring,

Reorganization of Flugparks into Fliegerkompanien as of 1 March 1915, Following War Orders	
***Flugpark* [Flight Park]**	***Fliegerkompanie* (*FliK*) [Flying Company]**
Flugpark № 1 - Wr. Neustadt	*Fliegerkompanie* 1 and 11
Flugpark № 2 - Wr. Neustadt	*Fliegerkompanie* 2 and 13
Flugpark № 3 - Görz	*Fliegerkompanie* 3 and 15
Flugpark № 4 - Mostar	*Fliegerkompanie* 4
Flugpark № 5 - Ujvidek	*Fliegerkompanie* 5 and 12
Flugpark № 6 - Mostar	*Fliegerkompanie* 6
Flugpark № 7 - Krakow	*Fliegerkompanie* 7 and 9
Flugpark № 8 - Przemysl	*Fliegerkompanie* 8
Flugpark № 10 - Graz	*Fliegerkompanie* 10
Flugpark № 14 - Aspern	*Fliegerkompanie* 14

Above: Before 1914, *Flugparks* [Flight Parks] operated as centralized hubs within the *k.u.k. Luftschifferabteilung*, supporting the Austro-Hungarian military's early aviation efforts. These facilities were responsible for maintaining, repairing, and storing the limited number of aeroplanes in service, as well as providing logistical support for training and experimental flights. With the outbreak of war in 1914, their role expanded significantly. On 1 March 1915, wartime orders reorganized the *Flugparks* into operational support centers for the newly established *Fliegerkompanien* (*FliK*) [Flying Companies]. Each *Flugpark* was tasked with supporting one or more *FliKs*, strategically positioned at key locations across the Austro-Hungarian Empire, such as Wiener Neustadt, Görz, Mostar, and Krakow. This transition marked a critical evolution from a pre-war logistical framework to an active operational support network, meeting the increasing demands of aerial reconnaissance, combat, and logistics during World War I. The accompanying chart details the allocation of *Fliegerkompanien* to their respective *Flugparks*, illustrating the distribution of Austro-Hungarian air assets during this transformative period.

flight theory, and military reconnaissance through a theoretical examination. A comprehensive manual was later released on 14 May 1914, outlining detailed instructions for various flight maneuvers, as well as useful guidelines for emergency procedures and medical fitness criteria.[32]

Between 1910 and the end of 1912, 88 *Feldpiloten* were trained, with an additional 66 in 1913, and only 11 more before the war began.[33] However, this total of 165 trained pilots does not accurately represent the available pilot resources for the *FliK*s in August 1914.

The Aeronautics Department faced a shortage of personnel and serviceable aeroplanes due to the ongoing buildup. Each of the 15 *FliK*s planned for mobilization required eight pilots and six aeroplanes, but only 85 pilots and 39 fully serviceable aeroplanes were available.[34] Comparing these figures to those of the German ally—which had approximately 230 serviceable aeroplanes for their 33 flying units and eight fortress flyer units in August 1914—reveals the extent of these shortcomings.[35]

Austrian sources and literature often downplay the true situation of the *k.u.k.* Aeronautics Department when the war began. In reality, the Danube Monarchy had the smallest air arm among all the Great Powers at war. The official history work *Österreich-Ungarns Letzter Krieg* [Austria-Hungary's last war] incorrectly claims that 15 aviation companies with a personnel strength of 106 men each were available and mobilized. In fact, only nine *FliK*s were sent to the Galician and Serbian fronts, with a mere seven assigned to the crucial northern front in Galicia.

Due to the earlier grounding of Lohner aeroplanes because of structural weaknesses, the number of serviceable aeroplanes was so low that the German ally had to assist with the delivery of Aviatik aeroplanes. In a hasty attempt to improve the grounded Lohner aeroplane inventory, only partial remedies were achieved. For example, *FliK 1* lost two of its Lohner *Pfeilflieger* during test flights without any enemy action. Various malfunctions occurred, such as breaking propellers and melting solder joints due to insufficient cooling. Another *Pfeilflieger* of *FliK 1* was downed by enemy rifle fire on 14 August because the aeroplane could not climb above 500 meters altitude over hostile territory.[36]

Above: A Lohner *Pfeilflieger*, with its distinctive swept-back wings, ascends over the Fischamend aviation complex, dominated by the imposing airship hangar with its twin towers in the foreground. This hangar housed and maintained airships during the early stages of *Luftschifferabteilung* development. Surrounding the hangar are various workshops, buildings, and testing facilities, reflecting the extensive industrial and technological operations at Fischamend. The complex played a critical role in aeroplane repair, maintenance, and development, as well as the training of military personnel, making it a cornerstone of the empire's military aviation efforts.

The documented personnel strength of 106 individuals per *FliK* appears to be inaccurate. For instance, *FliK 1* comprised ten pilot officers and a 50-member ground crew, while *FliK 7* had eight pilot officers and a 60-member ground crew. *FliK 10* was the only unit reported to have achieved the intended strength of 105 personnel and ten pilot officers.[37]

The aeroplanes, engines, and ground equipment were not the sole limitations. *Obstlt* Uzelac anticipated a shortage of young officers for aviation roles and expected significant losses in war. Consequently, he sought to train non-commissioned officers and enlisted men as pilots as well.[38] Despite resistance from conservative officer circles, Uzelac successfully broke traditional rank and station barriers by authorizing non-commissioned officers as well as enlisted men to become pilots.[39]

In addition to promoting the training of *Feldpiloten*, *Obstlt* Uzelac recognized the need for specialized training for aerial reconnaissance officers, as Germany had already initiated an extensive observer training program in January 1912.[40] Uzelac proposed establishing a training unit for *Beobachter* [Aerial Observer] officers at Hajmáskér, Hungary. Modeling after the German approach, he planned a four-week course for observer officers and a two-week orientation course for *Generalstab* officers to acquaint them with the potential and limitations of aeroplane deployment. While the proposal was insightful, practical implementation lagged, with the *Fliegeroffiziersschule* (*Flosch*) [Field Officer School] dedicated to training observers not being established until October 1915, reflecting broader challenges and bureaucratic delays within the Austro-Hungarian military structure.[41]

Despite Uzelac's foresight, he was unable to fully address the problems stemming from the severely understaffed and ill-equipped *k.u.k.* Aeronautics Department, as he lacked adequate time to achieve his objectives. When the war began, there was a significant deficit of reserve aircrews. Germany had used funds from its *Nationalflugspende*[42] to train pilots, resulting in a pool of 820 trained pilot officers, NCOs, and enlisted men by 1914. Although not all these individuals could be deployed in air units, approximately half were capable of wartime flying.[43] In contrast, Austro-Hungary had virtually no reserves, with only 15 civilian pilots requiring additional military training. The situation concerning *Beobachter* officers was similarly dire. The first training course for observer officers had yet to be completed when *FliK*s needed them,

Above: *Luftschifferabteilung Taube "ASRA"* with *Feldpiloten Oberleutnant* Robert Eyb and an observer, during preparations for a flight. The *Taube "ASRA"* was one of 14 Etrich 1912 *Taubes* (Type F) ordered by the *Luftschifferabteilung* from Lohner for the *Kaiserliche Herbstmanöver* [Imperial Autumn Maneuvers] of summer 1912. After initial flight tests, it was determined that the aeroplane was severely nose-heavy, necessitating a series of modifications to shift the center of gravity rearward. These changes included relocating the passenger to the rear, shifting the fuel tank, and moving the engine back by half its length. Each aeroplane in this series was christened with a unique name displayed on the fuselage. The names used were *AAR, ACHILLES, ADJUDANT, AFRA, AJAX, ALBUS, ALLONS, AMAZONE, AMOURETTE, ANITA, ANNY, ARAM, ARMIN, ASRA, ATAIR*. By 1915, surviving *Taubes* were redesignated as Etrich A.I(Fd) and primarily served as trainers within *Fliegerersatzkompanie* (*FleK*) [Aviation Replacement Company], powered by 65 hp or 90 hp Daimler engines or 85 hp Hiero engines. One machine remained in use as late as November 1916.

necessitating pilot officers to assume observer roles in the *FliK*s during the initial phase of the World War I.[44] In comparison, Germany had an astonishing 500 trained *Beobachter* officers at the war's outset.[45]

Endnotes

1. Peter, *Luftschiffer,* 70; Zelezny, *Neustadt*, 91.
2. Peter, *Luftschiffer,* 72, 74; Desoye, 93.
3. Peter, *Luftschiffer,* 74.
4. Desoye, 31.
5. Peter, *Luftschiffer,* 75; John Howard Morrow, *The Great War in the Air: Military Aviation from 1909 to 1921*, University of Alabama Press, Tuscaloosa, Ala., 2009, 25; Andrea Ungari, "The Italian Air Force from the Eve of the Libyan Conflict to the First World War," *War in History*, November 2010, Vol. 17, №. 4, 403–411.
6. Desoye, 48; Peter, *Luftschiffer,* 81.
7. Peter, *Luftschiffer,* 77.
8. Cisleithania, officially The Kingdoms and Lands Represented in the Imperial Council, (German: *Die im Reichsrat vertretenen Königreiche und Länder*) was the northern and western part of Austria-Hungary, the Dual Monarchy created in the Compromise of 1867—as distinguished from Transleithania (i.e., the Hungarian Lands of the Crown of Saint Stephen east of [beyond] the Leitha River). This name for the region was a common but unofficial one
9. The Lands of the Crown of Saint Stephen (Hungarian:

Above: The production of *Taube* aeroplanes at the *k.u.k. Fliegerarsenal Flugzeugwerk* in Fischamend during 1913. Initially serving the *Luftschifferabteilung* for airship and balloon operations, the facility was repurposed to meet the rising demand for military aeroplanes. By 1912, the repair shops were expanded into a full-fledged aeroplane manufacturing center, producing Etrich *Taubes* without proper licensing from the patent holders, Etrich and MLG, a move that reflected the increasing strain on Austro-Hungarian aviation resources. As a pivotal site for the construction, testing, and maintenance of military aeroplanes, Fischamend's importance to Austro-Hungarian aerial readiness grew as World War I loomed. Beyond aeroplane production, the facility also became responsible for a wide range of aeronautical advancements, including wireless communications, bombs, photography, and propeller wind tunnel testing. These developments established Fischamend as a cornerstone of Austro-Hungary's war effort, supporting diverse military aviation needs until 1918. *Credit: Christian Reiterer*

a Szent Korona Országai), informally Transleithania (meaning the lands or region beyond the Leitha River) were the Hungarian territories of Austria-Hungary, throughout the latter's entire existence (30 March 1867—16 November 1918), and which disintegrated following its dissolution. The name referenced the historic coronation crown of Hungary, known as the Crown of Saint Stephen of Hungary, which had a symbolic importance to the Kingdom of Hungary.

10. Conrad von Hötzendorf, *Aus meiner Dienstzeit, 1906-1918*, Band II, [From my service, 1906-1918], Rikola Verlag Wien 1922, 47.
11. Desoye, 37.
12. István Petróczy von Petrócz (3 January 1874– 9 August 1957) played a pioneering role in the development of military aviation within the Austro-Hungarian Empire. Breveted as a military aeronaut on 30 September 1902, he later became head of the *k.u.k. Luftschifferabteilung* (Imperial and Royal Aeronautics Department) at Fischamend on 2 February 1910. Achieving another milestone, Petróczy was breveted as the first *k.u.k.* military pilot from Hungary after passing his flight test on a Wright Flyer biplane on 22 July 1910. Under his command, a military flying school was established at Wiener Neustadt on 19 April 1911. As the commander of the *Flugarsenal* (Aviation Arsenal) in Fischamend, Petróczy was instrumental in the design and construction of the pioneering tethered helicopters PKZ-1 and PKZ-2 [Note: the term PKZ was invented after WWI; the period name was *Schraubenfesselflieger*]. Collaborating with *Oblt* Theodore von Kármán and Fähnrich Vilmos Zurovetz and utilizing the propeller designs of *Lt*Oszkár Asboth, his team tested prototypes that, while innovative, were ultimately deemed unstable for military use. Note on name variants: In German, which was one of the official languages of the Austro-Hungarian Empire, Petróczy von

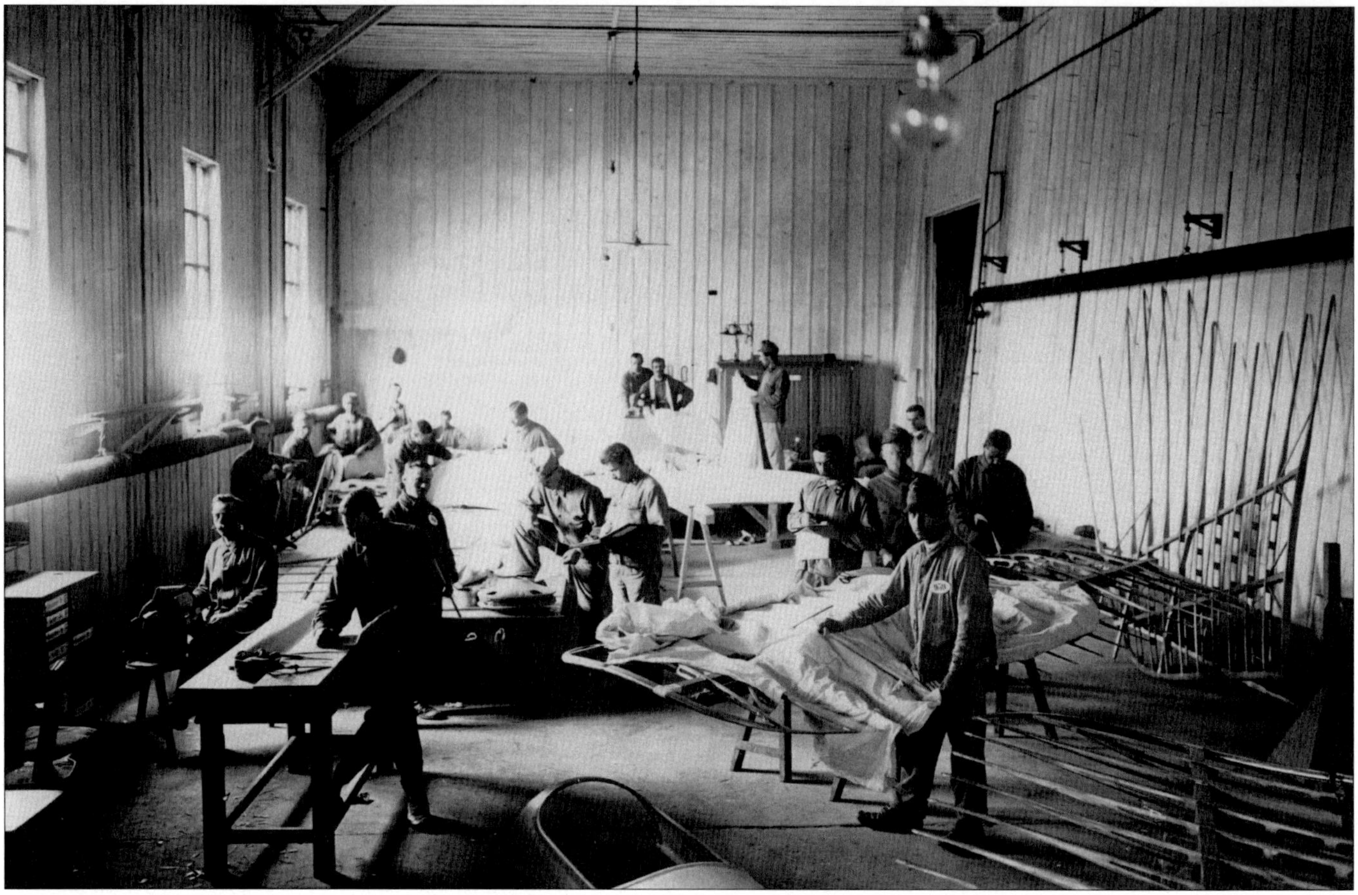

Above: A workshop space for the meticulous assembly of *Taube* wings at the *k.u.k. Fliegerarsenal Flugzeugwerk* in Fischamend highlights the facility's role in producing aeroplanes for the Austro-Hungarian military beginning in 1912. Unlike larger-scale private enterprises such as Lohner-Werke, the *Fliegerarsenal* functioned as a skilled workshop, where the production of individual aeroplanes required a variety of manual skills from its personnel. These included not only traditional woodworking but also precision metalworking, fabric stitching, and other specialized techniques essential to aeroplane manufacturing. The assembly of the *Taube's* wings exemplifies the craftsmanship and adaptability demanded in early aviation production. The War Ministry's reluctance to place larger orders for aeroplanes constrained production and limited the number of machines available as demand grew. This shortage compelled the *Fliegerarsenal Flugzeugwerk* to supplement Austro-Hungarian aviation needs by building *Taubes*. When aeroplanes became central to the *Luftschifferabteilung*, the facility expanded its operations beyond balloons and airships to include maintenance and repair work for aeroplanes. While its output remained modest, the facility played an important role in the development of Austro-Hungarian military aviation. *Credit: Christian Reiterer*

Petrócz's name would be rendered as "Stefan Petróczy von Petrócz." This German variant of his name highlights the linguistic diversity of the empire and may appear in German-language historical documents and references related to his work [Note: this was a general phenomenon with Hungarian names, e.g. Milan (Emil) Uzelac].

13. Under the command of *Obstlt* Emil Uzelac, the *k.u.k. Luftschifferabteilung* underwent significant reorganization, eventually leading to its transformation into the *k.u.k. Luftfahrtruppen* on 18 June 1915. Uzelac's leadership was instrumental in developing the *Luftschifferabteilung* into an effective air arm of the Austro-Hungarian military; Desoye, 44; Peter, *Luftschiffer,* 81.
14. After the war, Uzelac resided in the newly established State of Croatia and contributed to the development of the Croatian air service. During World War II, Croatia, an ally of Germany, appointed Uzelac as commander-in-chief of its Air Force: Desoye, 44. After 1945, Uzelac retired to Petrinje, then part of Yugoslavia and now Croatia, where he passed away on 7 January 1954.
15. Zelezny, *Neustadt*, 156-157: Peter, *Luftschiffer,* 80: *Das Offiziersfliegen* [Officers Flight Meeting], a monumental showcase of Austro-Hungarian military aviation, was held on 6 October 1912, at the Wiener Neustadt airfield. This event, part of the autumn maneuvers in southern Hungary earlier in September, featured military aeroplanes used for reconnaissance and reporting missions. Despite minimal coverage by the daily press on their successful use, the

Above: *Obstlt* Milan (Emil) Uzelac, commander of the *Luftfahrtabteilung*, stands prominently at the '*Das Offiziersfliegen*' [Officers' Flight Meeting] held on 6 October 1912 at Wiener Neustadt. Despite limited resources, Uzelac played a pivotal role in advancing the Austro-Hungarian aviation program, pressing forward with aviator training amidst a shortage of aeroplanes and equipment. In the background is the *Taube* '*ARAM*,' flown during the event by *Rittm* Hans *Ritter* Umlauff von Frankwell. *Credit: Christian Reiterer*

planning for this high-profile event began in August 1912, anticipating a large turnout. Indeed, the day saw over 25,000 spectators, including notables such as Archdukes Rainer and Karl Albrecht, Interior Minister Karl Heinold von Udinski, Finance Minister Dr. Waclaw von Zaleski, War Minister Moritz von Auffenberg, Navy Commander Admiral Count Rudolf Montecuccoli, and high-profile industry representatives like Camillo Castiglioni, Ferdinand Porsche, and Ludwig Lohner. The event displayed the prowess of newly purchased *F* series Etrich *Taube* aeroplanes and Lohner *Pfeilflieger*, performing various demonstrations. Among the aviators, *Oblt* Robert Eyb piloted the Etrich *Taube* named *Asra*, demonstrating the aeroplane's agility in challenging weather conditions. *Oblt* Rudolf Holeka, aboard the *Anita*, and *Lt* Josef Flassig, flying the *Adjutant*, showcased precision in navigation and flight control, critical for reconnaissance and operational success. The engineers, including Ferdinand Porsche and representatives from Daimler-Werke like Eduard Fischer, ensured that the aeroplanes were at the technological forefront, enhancing their performance under demanding conditions. Highlights of the day included a cross-country flight to Odenburg and precision landing demonstrations essential for pilot certification. The squadron flight was particularly memorable, with pilots navigating through gale-force winds, demonstrating complex aerial maneuvers, and concluding with a ceremonial flower drop over the audience. This demonstration of coordinated aerobatics and formation flying, led by seasoned pilots, tested the precision and effectiveness of group maneuvers, crucial for future military engagements. The event concluded successfully without incidents, raising 16,500 kroner ($3,300) in donations, showcasing the potential and precision of the *k.u.k. Luftschifferabteilung* capabilities, and significantly bolstering public and official support for the development of an Austro-Hungarian air fleet.

16. Peter, *Luftschiffer,* 86; In 1912, 7.9 million Austrian-Hungarian kronen would have been worth approximately 1,580,000 1912-USD; In 1912, 4.6 million Austrian-Hungarian kronen would have been worth approximately 920,000 1912-USD.
17. Peter, *Luftschiffer,* 86.
18. Desoye, 92: Jörg C. Steiner, „Das Feldpiloten-Abzeichen," *Militärhistorische Themenreihe*, Bd. 4. Wien: Edition S & H, 1992: Robert S. Pandis, "Flight Badges of the Central Powers, 1914-1918," Volume II—*The Imperial*

Above: Lohner *Pfeilflieger* B.I (Type B, Series H2, Military Series), powered by a 90 hp Austro-Daimler engine and featuring a distinctive rear wheel fitted with a band brake. The steel-tired rear wheel, equipped with protruding studs, was designed to enhance sliding resistance during braking, providing greater ground control on unprepared airfields. These two-seater aeroplanes, named *BLITZ, BABARA, BEATA*, and *BAJAZZO*, bore the red and white identification markings of the Austrian military, which were in use until the spring of 1915. *k.u.k. Luftschifferabteilung Pfeilfliegers* were prominently displayed and flown during a series of national events, including the Third Aspern Flight Meet (Das dritte Flugmeeting in Aspern), held from 21 to 29 June 1914. This celebrated event drew participants representing nations throughout Europe, showcasing cutting-edge aviation design and capabilities to both military and public audiences. While this aviation milestone unfolded, history took a darker turn. On 28 June 1914, Archduke Franz Ferdinand of Austria, heir presumptive to the Austro-Hungarian throne, and his wife, Sophie, Duchess of Hohenberg, were assassinated in Sarajevo. This pivotal event, occurring during the final days of the Aspern Flight Meet, set in motion a series of events that would lead to the outbreak of World War I. The juxtaposition of technological progress in aviation and the looming geopolitical crisis underscores the significance of this moment in history.

Above: The Lohner *Pfeilflieger* "*BÖMERANG*," a Type B reconnaissance biplane, illustrates the innovative yet transitional phase of Austro-Hungarian military aviation in 1913. Delivered to *Flugpark* 7 on 12 March 1913, it was powered by a 90 hp Austro-Daimler engine and featured a wingspan of 13.4 meters with a total wing area of 37.5 m^2, achieving a maximum speed of 115 km/h and a climb rate of 1,000 meters in 12 minutes. The "*BÖMERANG*" incorporated the "Spanish wing," a significant design refinement. The revised wing joined the upper halves directly at the centerline, eliminating the original cabane structure in favor of inverted-vee struts, which reduced weight and drag. Rectangular ailerons replaced the earlier rounded design, and folding wingtips allowed for easier storage—enhancements that improved practicality for field operations. These refinements, coupled with the distinctive double tail skid, made the "*BÖMERANG*" more efficient to maintain and operate from austere airfields. While praised for its docile flight characteristics, the Type B's structural limitations became apparent during operational use. Extensive wear and tear during the Balkan crisis highlighted these weaknesses, necessitating frequent repairs. Despite its shortcomings, the "*BÖMERANG*" and its contemporaries played an essential role in training and reconnaissance, supporting Austro-Hungarian pilots during a formative period in the nation's military aviation development. *Credit: Christian Reiterer*

Facing Page, Bottom: The Lohner *Pfeilflieger* B.I reconnaissance biplane, designated '*BARBARA*,' was stationed at the Zemun airfield near Belgrade (Beograd) in 1914. Housed in a typical field hangar, the aeroplane was protected from the elements and underwent preparation and maintenance tasks essential for reconnaissance missions. Distinctive in its appearance, the '*BARBARA*' features red-and-white markings on its wingtips, a visual identifier of Austro-Hungarian national identity, which also appeared on the fuselage. Strategically located near the Serbian front, the Zemun airfield played a pivotal role in reconnaissance and observation operations during the early stages of the conflict. While the early Lohner B.I models, built starting in 1912, showed promise during training and early operational service, including the Balkan crisis, production issues soon emerged. By 1914, the machines were found to be increasingly fragile and ill-suited for combat operations, with numerous aeroplanes grounded due to wing and fuselage failures. Despite these challenges, limited modifications were undertaken, including some attempts to improve wing strength, though these efforts ultimately relegated the B.I series to training roles by late 1914. *Credit: Christian Reiterer*

German Naval, Austro-Hungarian, Bulgarian, and Ottoman Turkish Air Services, Imperial House Antiques LLC,132.

19. Desoye, 37; In 1913, 5,500 Austrian-Hungarian kronen would have been worth approximately 1,100 1913 USD. In 1913, 17,500 Austrian-Hungarian kronen would have been worth approximately 3,500 1913-USD.
20. Desoye, 37: Prior to the Wiener Flugwoche (1st Flying Week) in Vienna in 1912, military pilots typically garnered lower rankings in competitions, which resulted in the public's predominant interest in civilian pilots. This perception shifted significantly after *Oblt* Philipp von Blaschke, serving as a military pilot, set two world records. Additionally, the organization of the officers' flying competition further catalyzed the rise in popularity of military aviation among the general populace.
21. Desoye, 49; In 1914, 1.4 million Austrian-Hungarian kronen would have been worth approximately 280,000 1914-USD.
22. Peter, *Luftschiffer*, 82.
23. Peter, *Luftschiffer*, 88
24. Alexander Löhr continued serving after 1938 in the German *Luftwaffe*. As a *Generaloberst*, he held high *Luftwaffe* commands during World War II on the German Eastern Front. Löhr was held responsible for the Belgrade bombardment of 6 April 1941. After the war, he was accused of war crimes by the Yugoslavs, sentenced to death, and executed on 26 February 1947 in Belgrade.
25. Because of the milder climate, they could also fly during the winter period, Görz now Novi Sad, Serbia; Peter, *Luftschiffer*, 89.
26 . Kahlen, *Entwicklung*, 3.
27. *Fliegerkompanien* [Flyer Companies]. The term was shortened to *FliKs*. These became the foundational flying unit in the *k.u.k Luftschifferabteilung* and later the *k.u.k Luftfahrtruppen*, typically with eight pilots. By war's end there would be a total of 77 *Fliks*.
28. Desoye, 44.
29. Kahlen, *Entwicklung*, 8; Desoye, 44.
30. St ahl, *Die Militärluftfahrt*, 184.
31. Desoye, 97.
32. Desoye, 63-86.
33. Desoye, 40.
34. Kahlen, *Entwicklung*, 1.
35. Stahl, *Die Militärluftfahrt*, 193.
36. Robert Veinfurter, *Das fliegende Personal der k.u.k. Fliegerkompagnien im Ersten Weltkrieg* [The Aviation personnel of the *k.u.k.* aviation companies in World War I], Volume 34 of ÖFH-Nachrichten, 19, 14 August 1914; Lohner *Pfeilflieger* Type *D* (B.III) *FliK 1* (2. Armee, Galizien) *Feldpilot Oblt* Hugo Schwab *Beobachter Oblt* Franz Kabelac. Russian troops at Balin, near Kameniec-Podolsk.
37. Meindl, *Luftsiege*, 3.
38. Peter, *Luftschiffer*, 94.
39. His upbringing, as a child of parents from a lower socioeconomic background, likely played a significant role in shaping his decision.
40. Stahl, *Die Militärluftfahrt*, 135; During World War I, the role of a *Beobachter* in the *k.u.k.* Aeronautics Department and later in the *k.u.k.* Air Troops was vital to the effectiveness of aerial operations. The primary responsibility of a *Beobachter* was to gather valuable intelligence and provide reconnaissance support from the air. Typically, they would fly in two-seater aeroplanes alongside a pilot who focused on navigating and flying the aeroplane, while the *Beobachter* operated cameras, utilized binoculars, and made detailed observations of enemy positions, movements, and infrastructure. In addition to reconnaissance, the *Beobachter* also played a crucial role in artillery spotting, helping to direct and correct the fire of ground-based artillery units by relaying information about the accuracy and impact of their shots. Furthermore, they sometimes operated machine guns or other onboard weapons for self-defense or to engage enemy aeroplanes. As the war progressed and aerial combat became more prevalent, the *Beobachter*'s role expanded to include additional duties such as navigating, assisting with aerial photography, and participating in bombing missions. Overall, the *Beobachter* was critical in ensuring the success of the *k.u.k.* Aeronautics Department and later the *k.u.k.* Air Troops missions throughout the conflict.
41. Peter, *Luftschiffer*, 91.
42. The *Nationalflugspende* [national flight donation] was a national fundraising campaign set up by the government before World War I to promote the development of the German aeroplane industry. At the suggestion of August Euler, Prince Heinrich of Prussia created the national flight donation in 1912 in the German Reich, based on the model of the zeppelin donation by the German people, in order to—as it was officially said—"unstoppably continue to work on the perfection of the flying machines, on training the aviator, so that dangers are reduced, and performance is increased." The fundraiser began with the prince's appeal for donations during his opening speech at the General Aviation Exhibition in Berlin, which took place from 3–14 April 1912. At the end of 1912, donations totaled 7.5 million marks; this would have been worth approximately 1.8 million 1912-USD.
43. Stahl, *Die Militärluftfahrt*, 142.
44. Meindl, *Luftsiege*, 5.
45. Stahl, *Die Militärluftfahrt*, 193.

8. Foundations of Austro-Hungarian Naval Aviation

Above: SMS *Budapest*, a Monarch-class coastal defense ship, photographed at the port of Fiume. On 16 July 1902, SMS *Budapest* became the first vessel in the *k.u.k. Kriegsmarine* [Imperial and Royal Navy] to conduct tests with an observation balloon. These early trials revealed significant challenges, such as the need to extinguish the ship's boilers during balloon ascents for safety reasons. Despite these initial efforts, integrating observation balloons into naval operations proved difficult, and the navy ultimately continued relying on traditional reconnaissance methods.

The Austro-Hungarian Empire, with vital maritime interests in the Adriatic and Mediterranean Seas, long recognized the importance of maintaining robust naval forces. Key ports like Trieste, Pola, and Fiume served not only as naval bases but also as vital economic gateways, connecting the empire to essential trade routes and ensuring the flow of resources. The historical experiences, such as the Battle of Lissa in 1866, underscored the kingdom's understanding of the critical importance of sustaining modern naval capabilities.[1]

By the late 19th century, the transition from sail to steam-powered propulsion marked a pivotal shift in naval capabilities, allowing ships to move independently of wind or current conditions and enhancing their speed and maneuverability. This evolution meant that ships could "move against the tide" both literally and figuratively, reshaping naval tactics and operations. With modern naval guns increasing the targeting distance to about 10,000 meters and ship speeds surpassing 20 knots, superior methods for targeting and reconnaissance became crucial. The line of sight from a ship was limited to around 5–8 kilometers for objects viewed from the deck and up to about 19–24 kilometers from the masthead. While the integration of telescopes in the 17th century revolutionized naval observation by enhancing visual acuity, it could not extend the inherent range dictated by the Earth's curvature. To overcome the limitations of ship-based observation, balloons were tested as observation platforms, extending the visual range beyond the age-old finite horizon.[2]

Above: Balloon test conducted between 1902 and 1907 by the *Ballonabteilung des Fünften Festungsartillerieregiments* [Balloon Detachment of the Fifth Fortress Artillery Regiment] of the *k.u.k. Armee*. This image captures a test conducted at Cattaro, where experiments assessed the feasibility of using observation balloons in coastal areas. Observers monitored ship formations, particularly the visibility of smoke plumes, under various weather and lighting conditions. Despite several years of experimentation, logistical challenges and budgetary constraints ultimately led the *Technische Komitee der Marine Sektion* [Technical Committee of the Navy Section] to deem captive balloons impractical for naval reconnaissance by 1906. Consequently, the *Kriegshafenkommando* [Naval Port Command] of Pola recommended focusing on optical signal stations instead. The final balloon exercises were completed in 1907.

Above: *Admiral* Rudolf Graf Montecuccoli, Chief of the Austro-Hungarian Navy from 1904 to 1913. Under his leadership in 1910, there was a renewed push for resources and guidelines to establish naval aviation. Montecuccoli emphasized the development of aerial reconnaissance capabilities to expand the navy's operational scope, modernizing the fleet ahead of World War I.

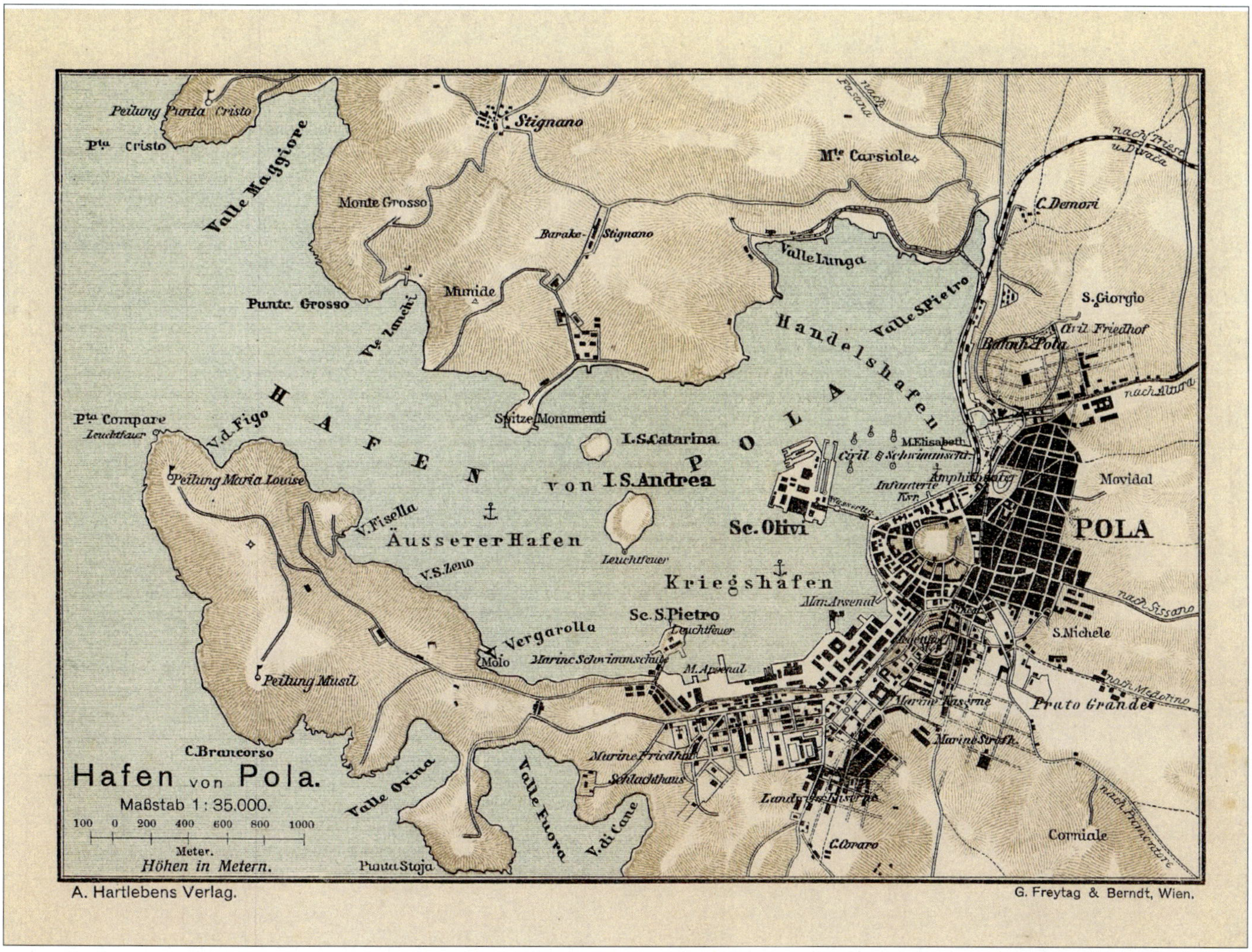

Above: Pola Naval Facility: Located on the Istrian Peninsula, Pola was one of the most important bases of the *k.u.k. Kriegsmarine* due to its strategic position along the Adriatic Sea, providing direct access to the Mediterranean. Its natural harbor made it an ideal location for a major naval base, and by the early 20th century, it had become the principal home of the Austro-Hungarian fleet. In June 1911, the *k.u.k. Kriegsmarine* launched an ambitious modernization program, focusing on constructing dreadnoughts and other large vessels. As part of this effort 100,000 Kronen were allocated specifically for the development of naval aviation. This initial funding, along with additional resources provided in June 1912, laid the foundation of the *k.u.k. Seeflugwesen* [Naval Aviation Service] station at Pola.

Initially, observation balloons were tested by various navies due to their availability, but numerous technical problems, such as dependence on favorable weather conditions and the dangers of hydrogen, limited their operational effectiveness. Ultimately, the need for an appropriate long-range scouting platform prompted a shift towards a new paradigm, setting the stage for naval aviation and the incorporation of dirigibles and aeroplanes for reconnaissance purposes.

Interest in aerostats within the *k.u.k. Kriegsmarine* [Imperial and Royal Navy] dates back to the mid-1880s. However, due to financial constraints, the navy initially limited its efforts to observing the balloon tests conducted by the *k.u.k. Armée* and foreign military powers. It wasn't until 16 July 1902 that the navy conducted its first test with an observation balloon aboard the pre-dreadnought battleship SMS *Budapest*.[3] These balloon ascents from the ship required extinguishing the ship's boiler fires for safety. Despite these early efforts, integrating observation balloons into naval operations presented substantial challenges. The exercises revealed that observation balloons, while useful in theory, were limited in practical application for various reasons, including dependence on favorable weather and difficulties in maintaining stability and control.[4] Consequently, the navy had no alternative but to continue relying on traditional observation methods for reconnaissance.

Above: *Vizeadmiral* [Vice Admiral] Juliusz Franciszek von Ripper, pivotal in leading the modernization of the *k.u.k. Kriegsmarine*, played a crucial role in advancing Austro-Hungarian naval aviation. As the driving force behind the establishment of the *k.u.k. Seeflugwesen* in 1911, von Ripper secured funds to develop essential naval aviation infrastructure. His vision for integrating aviation into naval strategy took shape at Pola, where the new aeronautical station trained officers and personnel in aerial reconnaissance and operations.

Above: *Linienschiffsleutnant* (*Lschlt*) [Lieutenant] Viktor Klobučar Rukavina von Bunic, an early pioneer of Austro-Hungarian naval aviation. On 12 August 1911, he was sent to the *Militäraviatische Station* [Military Aviation Station] in Wiener Neustadt to receive his *Feldpiloten* rating and become a breveted pilot and was awarded FAI pilot's certificate № 26 on 19 September 1911. He commanded the *Versuchflugstation* [Experimental Flight Station] at Pola's Altura field, under the authority of the *Marine-Technisches Komitee*. From autumn 1911 to 22 June 1914, he served as the Kommandant of the *k.u.k. Seeflugwesen* [Naval Aviation Service], overseeing significant improvements in training programs and addressing the shortage of aviation equipment.

In 1902, the first training *Ballonabteilung* [Balloon Unit] was established in Pola. This unit aimed to gain experience in using balloons in coastal areas and to familiarize navy personnel with the complexities of ballooning over the sea. Observers were tasked with assessing various operational parameters, such as the visibility of smoke plumes from different ship types in tactical formations under varying weather and lighting conditions, enhancing our understanding of the technical capabilities and limitations faced during these early tests.

Between 1902 and 1907, the frequent deployments of these units highlighted several logistical challenges. Notably, the need for numerous coastal stations and gas production facilities presented an untenable investment given the navy's persistent budgetary shortfalls. By 1906, the *Technische Komitee der Marine Sektion* [Naval Section Technical Committee] had concluded that captive balloons were not practical for naval reconnaissance, leading to their dismissal as a viable option for this role. Consequently, the

Above: *Lschlt* Heinrich (Huß) Huss was one of the first Austro-Hungarian naval officers to receive pilot training. He completed his military flight training at Wiener Neustadt in 1911, earning FAI pilot's certificate № 25 on 19 September 1911. Following his training, Huß was assigned to the *Versuchflugstation* [Experimental Flight Station] in Pola, where he gained essential skills in seaplane operations. On 24 December 1913, he was officially rated as a *Wasserflugzeugführer* [Seaplane Pilot].

Above: *Lschlt* Božidar Mažuranić, one of the first naval officers to undergo aviation training, completed his military pilot's training in 1911 at Wiener Neustadt, receiving his FAI pilot's certificate № 38 on 2 November 1911. Mažuranić's contribution to Austro-Hungarian naval aviation was part of the initial wave of officers trained to expand the navy's aviation capabilities.

Kriegshafenkommando [Naval Port Command] of Pola recommended enhancing optical signal stations instead of continuing with tethered balloons. Following this recommendation, no further balloon experiments were conducted within the *k.u.k. Kriegsmarine* post-1907.[5]

Despite this, 1907 saw some final balloon exercises, including mine-sweeping operations and the towing of submarine targets to depths of 20 meters. During these operations, chimney caps were fitted to the towing steamers for safety due to the combustibility of the hydrogen-filled balloon. Reflecting on his participation in these exercises, by 1909, *Linienschiffsleutnant* (*Lschlt*) [navy Lieutenant] Wenzel Leopold Wosеček criticized their naval application, citing numerous practical difficulties.[6]

The annexation of Bosnia and Herzegovina in 1908, which extended the Adriatic coastline under Austro-Hungarian control, spurred further development of naval bases and initiatives. This geopolitical change increased the importance of the Adriatic coast to the Empire, necessitating improved defense capabilities. Although there was no direct and immediate link to aviation initiatives at the time, the need for a greater observation range indirectly supported the eventual establishment of such a program. By 1910, under the leadership of *Admiral* Rudolf *Graf* Montecuccoli, there was a renewed push for resources and guidelines for naval aviation. His leadership emphasized the importance of developing aerial reconnaissance capabilities to enhance the navy's operational scope.[7]

Development and Expansion of Naval Aviation

Above: *Fregattenleutnant* (*Flglt*) [Lieutenant Junior Grade] Gottfried *Freiherr* von Banfield, one of Austro-Hungary's most renowned aviators, earned his FAI pilot's certificate № 67 on 15 August 1912. Known for his exceptional skills, Banfield later became a key figure in naval aviation, particularly during World War I.

Above: *Lschlt* Hugo Ockermüller completed his military pilot training at Wiener Neustadt in 1912, receiving his FAI pilot's certificate № 63 on 10 August 1912. He was part of the second group of naval officers trained as *Feldpiloten*, contributing to the expansion of the *k.u.k. Seeflugwesen.*

As the major seafaring powers entered the 20th century, the importance of naval aviation became increasingly apparent. Airships held promise and proved to be practical for naval forces. In this realm, the technological advancements made by Germany, along with the steadfast belief in the zeppelin, eventually led to their successful deployment. Germany's success with rigid airships was the exception to the general limitations experienced by other countries who attempted to incorporate this technology. Despite potential benefits for long-range reconnaissance, the naval administration dismissed airships primarily due to high costs, dependency on specific bases, and susceptibility to weather conditions, in addition to concerns about their visibility and vulnerability. Meanwhile, the concurrent development of the aeroplane presented a more efficient and economical means of achieving similar goals, thereby shifting universal attention to this emerging technology.

The introduction of aeroplanes marked an important technological advancement for Austria-Hungary. Initially driven by the potential of aerostats, both the army and the navy soon realized that aeroplanes could

Above: *Lschlt* Franz Mikuleczky completed his training at Wiener Neustadt in 1913, receiving his FAI pilot's certificate № 128 on 31 July 1913. As one of the third group of naval officers to receive aviation training, Mikuleczky helped further develop Austro-Hungarian naval aviation capabilities.

Above: *Lschlt* Gustav Klasing earned his FAI pilot's certificate № 130 on 13 August 1913. His training was part of the navy's broader effort to develop a skilled corps of naval aviators in preparation for future operations.

greatly enhance their reconnaissance capabilities. This realization led to the decision to incorporate aeroplanes into naval operations. Although the development of seaplanes initially faced various technological challenges, they eventually evolved into advanced models with innovative features, paving the way for more refined and effective naval aviation capabilities.

In June 1911, the *k.u.k. Kriegsmarine* embarked on an ambitious modernization program, primarily focusing on constructing dreadnoughts and other large vessels. As part of this effort, under the under the leadership of *Vizeadmiral* [Vice-Admiral] Juliusz von Ripper, it was decided to allocate 100,000 kronen specifically for the development of naval aviation, marking the beginning of the *k.u.k. Seeflugwesen* [Naval Aviation Service]. This initial allocation, along with additional funds provided in June 1912, laid the groundwork for establishing an aeronautical station in Pola to train naval officers and ground personnel in aviation operations.[8]

Although the primary financial target was shipbuilding, the broader funding allocation of 312.4 million kronen also created a robust fiscal foundation that would later facilitate investment in the nascent naval aviation program. Under the leadership of *Admiral* Montecuccoli, the navy's high command advocated for the integration of aeroplanes into naval operations and the establishment of dedicated aviation facilities. *Vizeadmiral* von Ripper, head of the Naval

Above: *Flglt* Glauko Prebanda earned his FAI pilot's certificate № 138 on 25 August 1913. Prebanda was among the early naval officers to undergo aviation training, contributing to the growing capabilities of Austro-Hungarian naval aviation.

Above: *Lschlt* Wenzel Leopold Wosecek received his FAI pilot's certificate № 65 on 12 August 1912 at Wiener Neustadt. He was part of the second group of naval officers trained as *Feldpiloten* and played a significant role in the development of naval aviation in the *k.u.k. Seeflugwesen*.

Section Technical Committee, played a crucial role in securing financial support for this emerging sector, enabling the purchase of seaplanes and equipment essential to the program's development.[9]

On 12 August 1911, *Lschlt* Viktor Klobučar Rukavina von Bunić, who had already received flight training in France, was sent to the *Militäraviatische Station* [Military Aviation Station] in Wiener Neustadt to become a breveted pilot by the Österreichischer Aëro-Club and to receive his *Feldpilot* rating. He successfully completed his military pilot's test on 12 August 1911 and was awarded his FAI pilot's certificate №. 26 on 19 September 1911. Klobučar obtained his seaplane pilot certification on 24 December 1913. To gain firsthand insight into the potential of aerial observation, *Admiral Graf* Montecuccoli joined him for a flight at the training airfield, underscoring the importance placed on aviation by the naval high command.[10]

Shortly afterward, *Lschlt* Klobučar was assigned command of the newly formed *Versuchflugstation* [Experimental Flight Station] at Pola's Altura field, which was under the jurisdiction of the Naval Technical Committee. The training programs were further expanded under his direction. From autumn 1911 until 22 June 1914, he served as *Kommandant* of the *k.u.k. Seeflugwesen*. His tenure was marked by noteworthy efforts to address the inadequate supply of flying equipment and to enhance the training and operational readiness of naval aviation personnel. Under his leadership, the naval aviation training programs saw considerable improvements, setting the foundation for the future expansion and effectiveness of Austro-

Right: L-R: Gottfried *Freiherr* von Banfield, Wenzel Leopold Wosеček, Konstantin Maglic. A 1915 photograph taken at *Seeflugstation* Kumbor, a key Austro-Hungarian naval aviation station on the Adriatic, captures a mix of naval officers and unrated seamen, illustrating the collaborative nature of operations at the facility. *Flglt* Gottfried *Freiherr* von Banfield, often referred to as the "Eagle of Trieste," was one of the Austro-Hungarian Navy's most accomplished flying aces and the final recipient of the *Militär-Maria-Theresien-Orden* [Military Order of Maria Theresa] for his service. *Lschlt* Wenzel Leopold Woseček, an early commander in the *k.u.k. Seeflugwesen*, was instrumental in shaping Austro-Hungarian naval aviation strategy and training programs. *Flglt* Konstantin Maglic contributed to reconnaissance and operational flights, emphasizing the critical role of seaplanes in Adriatic naval engagements. The presence of unrated seamen highlights the indispensable efforts of ground crews in maintaining and supporting early aviation endeavors.

Hungarian naval aviation.[11]

As naval aviation began to take shape within the *k.u.k. Seeflugwesen*, a methodical yet flexible approach was initially adopted to categorize and designate the various types of aeroplanes developed or acquired for maritime operations. Early experimental models and prototypes were classified under the provisional *Marineapparat* [Navy Apparatus] designation, reflecting their role in the nascent stages of the naval aviation program. However, by 1913, as the fleet expanded and the need for more specialized roles became apparent, a new system of designations was introduced, though these too were subject to change as the program matured. The *S* series was established to identify *Schwimmer* [floatplanes], crucial for operations at sea, and was later extended to include flying boats, which had proven indispensable for reconnaissance and patrol duties. Additionally, the *E* series emerged, attributed to the collaborative efforts of Igo Etrich and the team he worked with, signifying the Lohner-developed flying boats that incorporated their combined design principles. The *Seearsenal* [Navy Arsenal] at Pola contributed to this system with the *A* designation for aeroplanes built domestically, tailored specifically for the navy's training and operational needs. Finally, the *L* designation was reserved for special models, often prototypes or unique designs, signifying their experimental or limited production status. This structured yet evolving nomenclature underscored the strategic approach of the *k.u.k. Seeflugwesen*, ensuring that each type of aeroplane was effectively integrated into the broader objectives of naval operations.

In mid-1911, the planned training of naval officers as *Feldpilot* commenced with *Lschlt* Heinrich Huss and *Lschlt* Božidar Mažuranić being sent to Wiener Neustadt for military pilot training alongside army officers. Huss completed his training and received his FAI certificate 25 on 19 September 1911.[12] Mažuranić completed his training and received his FAI certificate 38 on 2 November 1911. Huss was then assigned to the Experimental Flight Station in Pola for a subsequent six-month assignment. This phase was crucial as it provided Huss with the opportunity to develop the skill of required for seaplane training. On 24 December 1913, his dedication and proficiency were recognized with his rating as a *Wasserflugzeugführer* [seaplane pilot]. This would enable the naval service to utilize his expertise by appointing him as a flight instructor, placing him at the forefront of training naval aviators.

The next group to complete the *Feldpilot* training at Wiener Neustadt was in mid-1912, consisting of *Lschlt* Robert Teufl von Fernland, brevet №. 57 on 27 July 1912; *Lschlt* Hugo Ockermüller, brevet №. 63 on 10 August 1912; *Lschlt* Wenzel Leopold Woseček,

Left: Josef Mickl, an innovative naval and aviation engineer, played a crucial role in the development of Austro-Hungarian naval aviation. In 1910, he was tasked with designing the empire's first seaplane, overcoming significant technical challenges with limited resources at the Pola naval arsenal. Mickl's groundbreaking efforts led to the production of the *Marineapparat I* in 1912, marking the launch of the Austro-Hungarian seaplane program. His work focused heavily on the development of floats and seaplane hulls, contributions he continued to refine throughout World War I.

Below: *Marineapparat I* undergoing final engine adjustments at the Pola naval arsenal in preparation for static thrust tests. The image shows the modified biplane configuration with its central float, designed by Josef Mickl after earlier monoplane attempts failed. Notably, the engine is mounted at the front of the fuselage with a chain drive connected to the elevated propeller, an engineering solution devised to prevent water contact during takeoff and landing. This phase of testing was essential to address the aeroplane's cooling and hydrodynamic issues, which delayed its full operational deployment.

Above: *Marineapparat I* during its sea trials near St. Catarina Island in early 1912. Despite earlier difficulties with water takeoffs due to the lack of a "step" on its central float, the seaplane was finally modified with twin-floats, improving its stability. This configuration enabled the aeroplane to successfully complete water operations and mark the beginning of Austro-Hungarian naval aviation's operational capabilities.

brevet №. 65 on 12 August 1912; and *Fregattenleutnant* (*Frglt*) [navy Ensign] Gottfried *Freiherr* von Banfield, brevet №. 67 on 15 August 1912. In early 1913, the third group of naval officers underwent training as *Feldpiloten* at Wiener Neustadt. This included *Lschlt* Franz Mikuleczky, who received his brevet №. 128 on 31 July 1913; *Lschlt* Gustav Klasing, brevet №. 130 on 13 August 1913; *Lschlt* Hugo von Wiktorin, brevet №. 133 on 19 August 1913; *Frglt* Karl Cerri, brevet №. 136 on 25 August 1913; *Frglt* Konstantin Maglić, brevet №. 137 on 25 August 1913; *Frglt* Glauko Prebanda, brevet №. 138 on 25 August 1913; *Frglt* Otto *Ritter* Fiala von Fernbrugg, brevet №. 140 on 27 August 1913; and *Frglt* Alois Kaindl, brevet №. 141 on 29 August 1913. These officers played key roles in the continued development of the *k.u.k. Seeflugwesen*, with their training reflecting the navy's commitment to expanding its aviation capabilities.[13]

At the beginning of 1910, the *k.u.k. Kriegsmarine* decided to embark on the construction of a seaplane within the naval arsenal, selecting *Diplom-Ingenieur* [Graduate Engineer] Josef Mickl as the designer. Mickl, who had joined the navy after completing his formal engineering studies, was known for his keen interest in aviation. Assigned to the arsenal in Pola, he was provided with a team of eight naval personnel. Despite limited resources, Mickl undertook the challenge of designing the aeroplane's wing structure based on the available information. He encountered various technical difficulties, particularly with the engine and its cooling system. To address these challenges, Mickl conducted engine tests, enabling him to measure both static and dynamic thrust, and devised methods to improve cooling.

Initially designed by Mickl as a landplane in a monoplane configuration, the aeroplane that would later be designated *Marineapparat I* underwent significant modifications after the original design proved unsuccessful. Completed as a biplane in the spring of 1911, the aeroplane was soon adapted for water operations with the addition of a central plywood float. However, achieving successful flight results as a floatplane took over a year due to challenges with hydrodynamic performance. With design input from

Above: *Marineapparat I* positioned on the slipway during testing. The central plywood float, along with a twin-float outboard arrangement, is clearly visible, showcasing the aeroplane's innovative design aimed at overcoming the challenges of water operations. The twin-float configuration, implemented after initial difficulties, provided greater stability during takeoff. The first successful water flight took place on 27 September 1912, with *Lschlt* Viktor Klobučar at the controls, covering a distance of 600 meters.

Lschlt Viktor Klobučar, the configuration was further refined, eventually incorporating two smaller floats attached to each wing, which improved stability. The first flight of the *Marineapparat I* took place on 29 July 1911 on Altura Heath. Powered by a 40-hp Austro-Daimler engine with a clutch and starter, and equipped with a three-bladed aluminum propeller, this flight marked the aeroplane's initial success, though water operations proved more challenging.

In early 1912, the *Marineapparat I* was transported from the naval arsenal to the island of St. Caterina for further testing on water. Despite continued efforts, the aeroplane struggled to take off from the water due to the absence of a "step" on the central float, which prolonged its takeoff run. After a complete redesign to a twin-float configuration, the *Marineapparat I* successfully took off from the water on 27 September 1912, with *Lschlt* Klobučar at the controls, reaching an altitude of 6–8 meters and covering a distance of 600 meters. By December 1912, the aeroplane completed its most extended flight to date, successfully flying a circuit around the island of Brioni, covering approximately 15–20 kilometers. Continued test flights led to further refinements, and the *Marineapparat I* served as a training machine for two years before being decommissioned in September 1914.[14]

Early in 1912, the *k.u.k. Kriegsmarine* acquired a Paulhan-Curtiss floatplane that belonged to *Lschlt* Viktor Klobučar, which he had initially purchased for personal use. The aeroplane was damaged during rail transport to Pola, leading Klobučar, who lacked the funds for the necessary extensive repairs, to generously donate it to the naval service. The floatplane was subsequently repaired and renamed *Marineapparat II.* Originally equipped with an 85-hp Hiero engine, the aeroplane faced persistent issues with underperformance and overheating. Despite efforts to address these problems by installing a 65-hp Austro-Daimler engine, *Marineapparat II* never achieved successful flight. The aeroplane was eventually

Above: The Paulhan-Curtiss Triad floatplane was built under license in France by Louis Paulhan, following the original Curtiss design without major modifications. This early seaplane was notable for its time, playing a significant role in the development of hydro aviation.

dismantled in early 1914, with its fuselage repurposed for experimental use.[15]

This was followed by the commissioning of *Marineapparat III*, a Lohner *Pfeilflieger* that was officially accepted at the Wiener Neustadt airfield on 24 April 1912. Initially built as a landplane, this model was first tested at Wiener Neustadt, where it demonstrated strong performance, achieving a full weight of 1,134 kg and reaching an altitude of 2,337 meters at a speed of 82.4 km/h. The landplane tests were conducted by *Lschlt* Philipp von Blaschke, with *Lschlt* Božidar Mažuranić and *Frglt* Heinrich Huss, all who had been trained at Wiener Neustadt and participated in the acceptance flights before its conversion to naval use with floats in August 1912. The aeroplane arrived in Pola on 23 May 1912, where it underwent unsuccessful experiments with the floats, which initially included two long, narrow runners connected by a horizontal bar mounted under the *Pfeilflieger* that were intended to help the aeroplane glide over the water's surface during takeoff and landing, and later removed. Despite these challenges, the machine finally managed to take off from the water surface in August 1912. It continued in service, primarily used as a trainer until September 1914. Powered by a 120-hp Austro-Daimler six-cylinder inline engine, the aeroplane had a wingspan of 16.80 meters and a length of 9.70 meters. Its empty weight was 540 kg, which increased to 800 kg once equipped with floats, reflecting the necessary modifications for marine operations.[16]

The Warchalowski Autoplan *Hydroplan* Type XI, designed in 1911 to meet the requirements of the *k.u.k. Kriegsmarine*, was a pusher-design, two-seat, open-frame, twin-float naval biplane. Power was supplied by an 85- to 90-hp Werner & Pfleiderer Hiero four-cylinder inline water-cooled engine, which drove a two-blade pusher propeller through a system of transmission shafts and gears. The overall dimensions included a wingspan of 15 meters and a length of 10 meters. Empty and maximum loaded weights were 650 kg and 910 kg, respectively.

The aeroplane made its inaugural flight on 25 June 1912, with Adolf Warchalowski at the controls. During flight tests in early August 1912, the Type XI completed a series of flights lasting as long as 30 minutes at various altitudes with different passengers, including *Ing* Karl Sehsun, *Frglt* von Banfield, and *Lschlt* Božidar Mažuranić. On 10 August 1912, during

Above: Following the success of earlier designs, the Navy expressed interest in a *Pfeilflieger* aeroplanes. After the *Pfeilflieger* was damaged in a flight competition in 1911, engineer Karl Paulal modified the aeroplane to meet naval requirements. The key modification involved raising the propeller shaft to accommodate floats, which was achieved using a chain drive from the engine to the raised propeller. These modifications were first tested on the Lohner-Daimler *Pfeilflieger II*, proving the viability of the design. The positive results led to the development of the Lohner-Daimler *Pfeilflieger III* for naval use.

the acceptance flight, the Type XI carried a 260 kg load, including an observer *Lschlt*. Wenzel Leopold Woseček, fuel, and ballast to simulate full operational capacity. During this flight, it remained airborne for over an hour, reached an altitude of up to 700 meters, and achieved a speed of 70 km/h.[17]

Following these successful land-based trials, the Type XI was delivered to the naval base at Pola and officially designated *Marineapparat IV*. There, it underwent tests with floats. Initially, the float system comprised two main flat-bottomed wooden floats supported on a metal-tube framework and a small single float under the tail. However, these floats were poorly designed, being almost as wide as they were long, which offered too much resistance and compromised stability.

On 29 September 1912, during a takeoff attempt, the aeroplane capsized, leading to a significant crash. The floats were sent back to Warchalowski for redesign. In response, a new single-float arrangement was constructed at the naval arsenal, which enabled the aeroplane to finally take off from the water in early 1913. Despite these modifications, *Marineapparat IV's* operational career was short-lived. On 21 April 1914, the aeroplane was involved in another crash, which damaged it beyond repair, leading to its subsequent write-off.[18]

In parallel, the navy looked to Lohner to further its seaplane capabilities. On 22 August 1912, the order for *Seeflugzeug Marineapparat V* was placed, followed by the order for *Marineapparat VI* on 25 September 1912. These biplane flying boats, commissioned with specific requirements for range, speed, and passenger capacity, were initially planned to be delivered with landing gear for the testing phase. Construction commenced in November 1912, with expectations high that these flying boats would bolster the navy's aerial capabilities.

However, the development of these aeroplanes was anything but smooth. Delays in the delivery of the essential 120-hp Daimler engines meant that construction timelines were compromised, and the aeroplanes faced a series of challenges during testing. *Lschlt* Viktor Klobučar undertook the acceptance flight of *V* in Pola on 25 February 1913, but the flight ended disastrously when the central float failed upon landing, causing serious injury to Klobučar. This incident necessitated a return to the drawing board, with Lohner's engineer, *Ing* Paulal, making critical modifications, including the addition of a "step" to the

Above: Lohner *Marineapparat III* prior to float installation. Initially built as a landplane, this Lohner *Pfeilflieger* was tested at Wiener-Neustadt in April 1912, demonstrating strong performance with a full weight of 1,134 kg, reaching an altitude of 2,337 meters at 82.4 km/h. After conversion for naval use with floats, the aeroplane underwent testing in Pola in May 1912, overcoming challenges related to water takeoff. Powered by a 120 hp Austro-Daimler engine, it was used primarily as a trainer until September 1914.

float design.

The troubles continued with *VI*. Although it was presented for acceptance in May 1913, the central float once again proved problematic, hindering the aeroplane's ability to take off from the water. Further modifications followed, but the aeroplane's issues were far from resolved; during a subsequent test in June, the propeller exploded. It wasn't until 28 July 1913 that the aeroplane finally completed a successful flight. Despite staying airborne for over an hour and reaching an altitude of 1,500 meters, its poor flying characteristics rendered it unsuitable for its intended training role. Consequently, *VI* was decommissioned in September 1914, after various attempts to rectify its flaws, and it was redesignated as *S.6*.

Adding to the Navy's frustrations were suspicions that Lohner had secretly swapped the aeroplanes *V* and *VI* without notifying the naval arsenal. A thorough investigation revealed that after the accident with *V*, Lohner had deceptively presented it as *VI* during the acceptance trials in May 1913 because the original *VI* had become unusable. This incident severely strained the previously strong relationship between Lohner and the Navy. In response, the Navy ordered a new seaplane in March 1914, designated as *L.5*, equipped with a 150-hp six-cylinder Austro-Daimler engine, to compensate for the deficiencies. Unfortunately, when the flying boat *S.5* (which was actually the problematic *V*) was delivered in May 1914, it was found to be obsolete and was rejected by the Navy. Lohner, facing mounting difficulties, eventually donated *S.5* to the naval arsenal in August 1914, where it was scrapped the following month.[19]

In November 1912, the Navy Arsenal in Pola constructed another floatplane under the direction of *Marineingenieur* Josef Mickl. This aeroplane, designated *Schwimmflugzeug Arsenal S.7*, had its maiden flight on 12 September 1913. However, due

Above: Warchalowski Autoplan *Hydroplan* Type XI - *Marineapparat IV*. Designed in 1911, this pusher-configured, two-seat naval biplane was powered by an 85-90 hp Werner & Pfleiderer Hiero engine. With a wingspan of 15 meters and a length of 10 meters, the aeroplane made its first flight on 25 June 1912. After successful flight trials, it was delivered to Pola as *Marineapparat IV* for water testing. Despite initial challenges with its float design, it managed a few successful flights but was ultimately written off following a crash on 21 April 1914.

to significant technical issues, it was not approved for service until 1915. Intended as a training machine, the *S.7* was powered by an 85-hp Hiero engine, but its underpowered engine and challenging flight characteristics severely limited its operational use. The aeroplane was stationed at St. Caterina and was eventually decommissioned in June 1915.[20]

After the initial experimentation, the *k.u.k. Kriegsmarine*, recognizing the necessity for operationally proven seaplanes, opted to procure Donnet-Lévêque flying boats from France—despite ongoing political tensions. These flying boats, equipped with Gnome rotary engines, were purchased by the *k.u.k. Seeflugwesen* and were instrumental in advancing training and reconnaissance operations, exemplifying a significant shift in naval aviation strategies. The first Donnet-Lévêque flying boats were delivered in early 1913, with subsequent orders made as the navy sought to expand its fleet rapidly.[21]

By 1912, with *Admiral* Montecuccoli's support, the first Experimental Flight Station was relocated from the Pola-Altura airfield to St. Caterina Island on 23 February 1912. This move included the construction of comprehensive aviation facilities, featuring six hangars capable of housing up to 20 aeroplanes, along with runways, fuel storage, and wharfs, thus supporting the operation and maintenance of seaplanes. This development reflected an important investment in naval aviation infrastructure. One of the important tasks undertaken was the development of removable ramps to facilitate the takeoff of floatplanes and flying boats from the limited deck space of warships, enabling the *k.u.k. Kriegsmarine* to extend its reconnaissance capabilities at sea.[22]

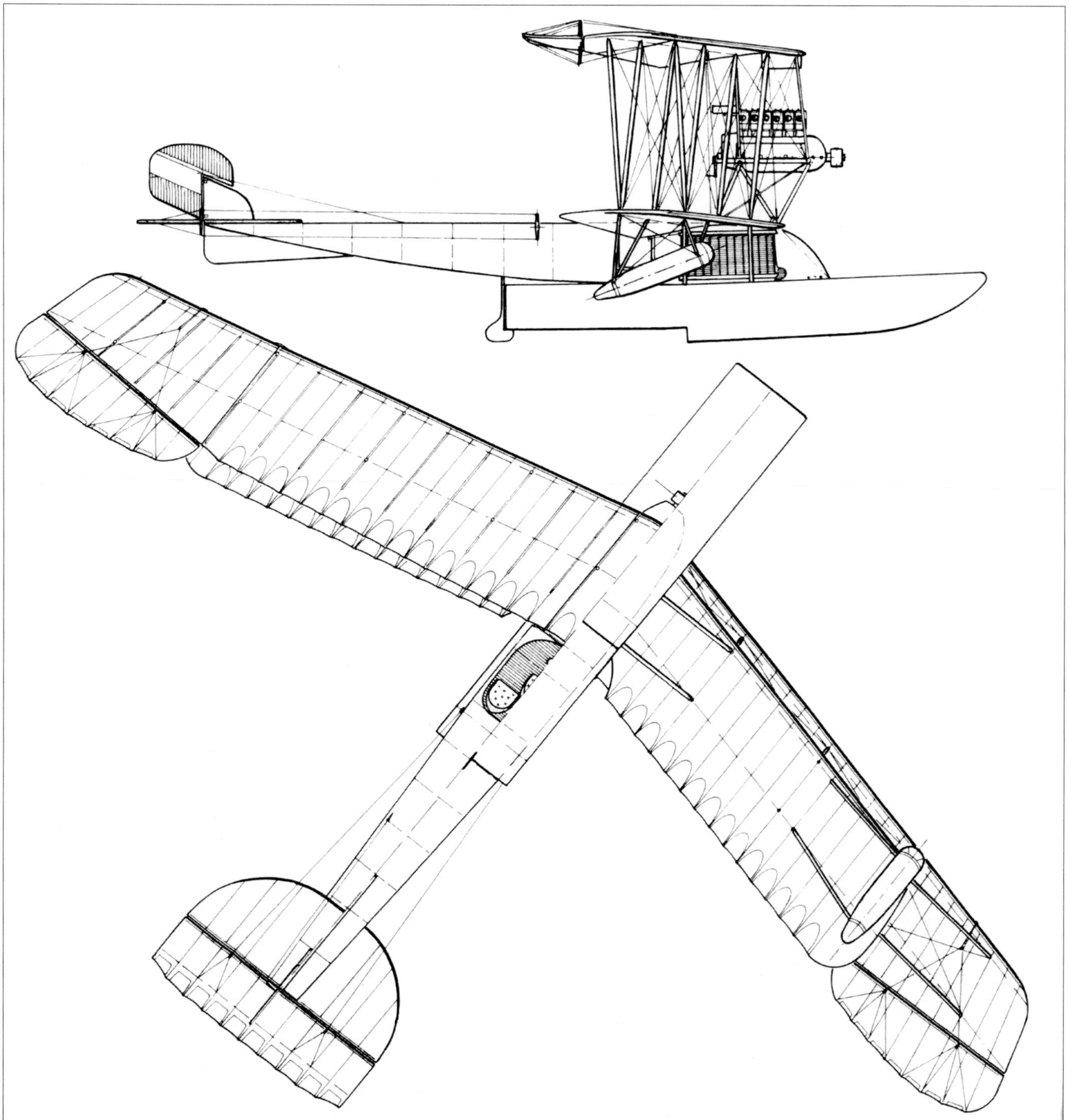

Above: Lohner *Marineapparat V*. Ordered on 22 August 1912, this biplane flying boat was intended to enhance the Navy's seaplane capabilities. Delays in receiving the 120 hp Daimler engine pushed back testing. On 25 February 1913, *Lschlt* Viktor Klobučar conducted the acceptance flight in Pola, but a failure of the central float upon landing resulted in his serious injury, prompting a redesign that included the addition of a "step" to improve hydrodynamic performance.

The concept of utilizing rails and platforms on ships for launching seaplanes was emblematic of the *k.u.k. Kriegsmarine* early forays into integrating aviation with naval operations. By 1911, as plans were laid to relocate operations from Altura Field to the strategically significant St. Caterina Island, the naval high command recognized the need to explore innovative solutions to expand their operational reach. The decision to install launch rails on ships such as SMS *Erzherzog Karl* and SMS *Gäa* reflected a forward-thinking approach

Above: R-L Three Donnet-Lévêque and the Lohner *Marineapparat V* (far end). After early experiments in seaplane development, the *k.u.k. Kriegsmarine* turned to French-built Donnet-Lévêque flying boats to meet its operational needs. These Gnome rotary engine-powered seaplanes were crucial in improving both training and reconnaissance efforts. The first Donnet-Lévêque flying boats arrived in early 1913, with further acquisitions reflecting the Austro-Hungarian navy's commitment to expanding its fleet and enhancing its aviation capabilities.

that sought to harness the potential of seaplanes for reconnaissance and combat support.

However, the practical challenges soon became apparent. The proposed installation of downward-sloping rails on SMS *Erzherzog Karl*, intended to run from the bridge over the front turret group, presented a significant tactical dilemma. The design, while theoretically sound, would have severely restricted the mobility of the ship's main guns, compromising its combat effectiveness. This realization, coupled with the logistical complexities of retrofitting existing ships with such launch mechanisms, led to a reevaluation of the approach. By August 1912, the idea had evolved into a proposal for a removable steel cable runway on SMS *Gäa*, but the cumbersome nature of the equipment and the operational difficulties it posed ultimately led to its abandonment.[23]

These early experiments, though not successful, were crucial in shaping the navy's understanding of the limitations and potential of ship-launched aviation. The notion of large landing decks on ships was briefly considered in April 1912 but was swiftly discarded due to the impracticality of dedicating entire vessels to this purpose, especially within the constraints of the Austro-Hungarian naval budget and fleet composition. The focus, therefore, shifted toward procuring and developing seaplanes that could operate independently from shore-based facilities, a decision that was further influenced by the strategic imperatives of the time.

Overall, this period was marked by a substantial investment in both the technological development of seaplanes and the training of personnel, laying the foundation for the effective deployment of naval aviation within the *k.u.k. Kriegsmarine*. During this period, the navy also prioritized the training of personnel. The initial training programs for the *k.u.k. Kriegsmarine*'s first officers in piloting and aviation technology were not only privately funded but also complemented by government-supported initiatives. Training was conducted at air schools in France and later at the *Militäraviatische Anstalt* [Military Aviation Facility] in Wiener Neustadt. *Lschlt* Božidar Mažuranić and *Frglt* Heinrich Huss were among the early naval officers to receive comprehensive training, reflecting a strategic emphasis on developing domestic training capabilities. This decision was driven by the need to conserve the limited number of seaplanes, which were essential for operational use and could not be spared for training. The use of land planes for initial training was not merely a matter of resource management but

Above: St. Catarina Island. The relocation of the *Versuchflugstation* to St. Catarina Island in 1912 was driven by the need for a more strategic and accessible base to support Austro-Hungarian naval aviation. Situated within the naval stronghold of Pola, St. Catarina offered proximity to the sea and ample space for expanding facilities, including hangars and wharfs for seaplanes. This move reflected the *k.u.k. Kriegsmarine's* recognition of the island's value for facilitating seaplane operations and experimentation.

also part of a deliberate strategy to build a robust pilot training program. Training initially on land planes allowed pilots to develop essential takeoff and landing skills on solid ground before transitioning to the more specialized and demanding tasks of operating seaplanes, thereby ensuring a more efficient and effective training process.[24]

In 1913, the *k.u.k. Seeflugwesen* formulated an extensive plan to acquire a large number of seaplanes, strategically allocated to support the naval fleet as well as coastal stations. The plan also included provisions for reserve aeroplanes. Despite these ambitious goals, the expansion faced a variety of obstacles, mainly due to a strong preference for sourcing from domestic manufacturers, which unfortunately lacked the capacity to meet the navy's then pressing demands.

In 1913, the year marked a pivotal one for Austro-Hungarian naval aviation, with major financial commitments and planning. *Admiral* Montecuccoli presented a comprehensive budget request that included a three-year supplementary funding of 170 million kronen, with a portion explicitly allocated for naval aviation. A supplementary expenditure of 205 million kronen was approved, with 484,000 kronen specifically earmarked for naval aviation, reflecting its growing importance. These funds were instrumental in acquiring new seaplanes, establishing training programs, and building the necessary infrastructure to support aviation operations. The budget reflected the navy's commitment to enhancing its reconnaissance capabilities through the integration of aviation technology.[25]

By early 1914, the *k.u.k. Kriegsmarine* continued to expand its naval aviation capabilities. This included the development of retractable wings, shooting trials with repeating pistols, and lighting tests on aeroplanes. On 21 February 1913, the first test on the operational capability of seaplanes in wartime took place with the flying boats 10 and 11 testing mine detection from the air. The results were excellent, and during World War I, mine detection became one of the main tasks of the *k.u.k. Seeflieger* [navy pilot]. The first bomb dropped

Above: SMS *Erzherzog Karl*, the lead ship of the *Erzherzog Karl*-class pre-dreadnought battleships, played a pivotal role in early Austro-Hungarian naval aviation experiments. In 1911, the ship was part of a forward-thinking initiative to integrate seaplane operations by installing removable ramps for launching aeroplanes from its deck. However, the practical challenges of interfering with the main guns' mobility led to the abandonment of this idea. These early trials shaped the navy's approach to aviation, ultimately leading to a focus on shore-based seaplane operations.

from a *k.u.k.* seaplane occurred on 15 August 1914, by *E.18* over Krstac, where the bombs had to be manually thrown at the target by the observer from the cockpit.[26]

By 1914, the *k.u.k. Kriegsmarine* continued to prioritize naval aviation, significantly increasing its budget allocation. Almost three million kronen were dedicated to expanding naval aviation, reflecting the program's growing importance within the overall naval operations. This budget supported the ongoing construction of aviation infrastructure and aeroplane acquisition, laying the groundwork for enhanced operational capabilities. Total naval expenditures for January to June 1914 included 134 million kronen for various naval projects, with a portion directed towards naval aviation. The plans for 1914 aimed to expand the number of operational aeroplanes from 15 to 80 by 1919, demonstrating the navy's ambitious vision for its aviation program. This expansion included the construction of new aviation facilities, such as the training centers at Experimental Flight Station St. Caterina and Training Flight Station Kosada, which were highly active in preparing personnel for military duties. By 1914, reflecting a growing commitment to this new military domain, the naval aviation budget had appreciably increased. Initially reluctant to utilize aeroplanes, the navy eventually recognized their value for extensive reconnaissance missions, marking a pivotal shift in naval policy.[27]

By the summer of 1914, Pola's inventory included approximately 22 seaplanes, with a significant portion dedicated to training purposes. Located along that part of the Adriatic Coast under control of the Dual Monarchy, these extended from near Trieste in the north to Durazzo in occupied Albania at the south. Prior to the war, there were only three such stations: the headquarters at Pola, the training center at nearby Kosada, and the operating base at Kumbor in the Bay of Kotor. A new permanent station was also established at Kotor during this period. This expansion reflected the importance of naval aviation and its integration into the broader military operations of the *k.u.k. Kriegsmarine*.[28]

The *k.u.k. Kriegsmarine*'s officer corps and enlisted

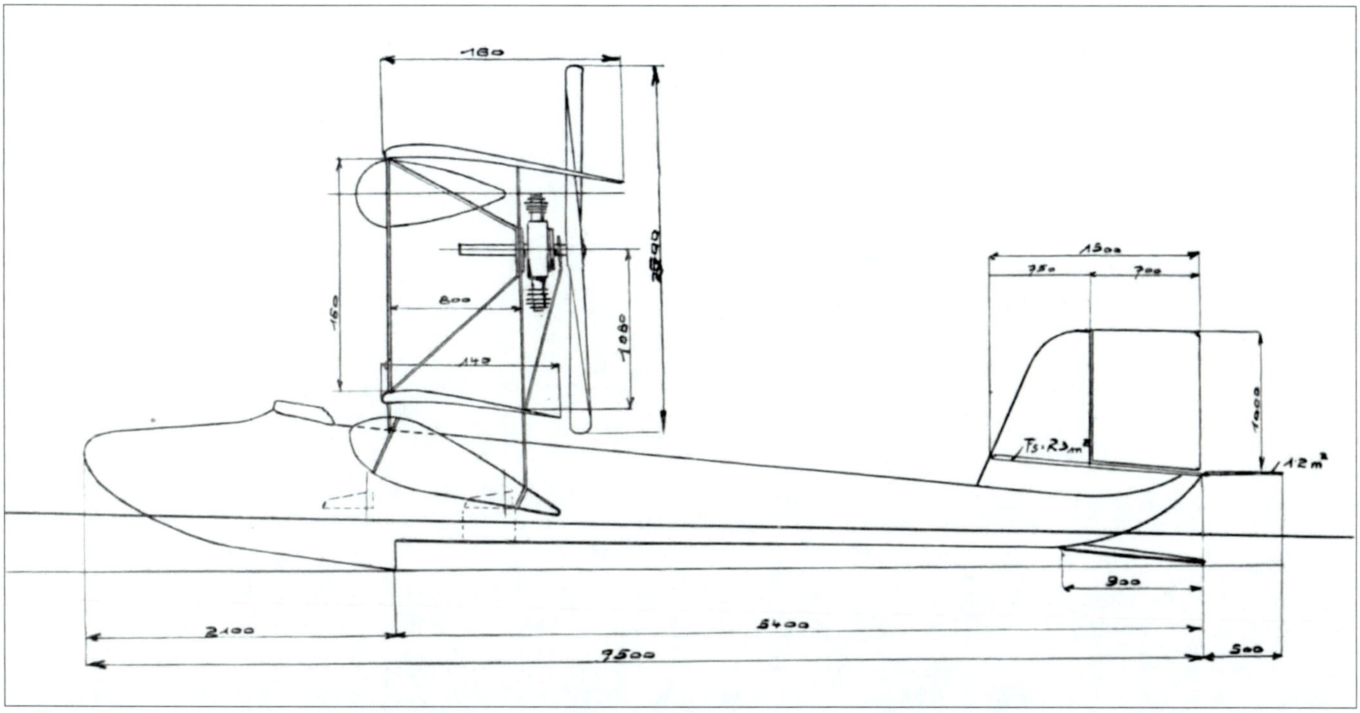

Above: St. Josef Mickl's detailed drawing of the Donnet-Lévêque flying boat, published in *Mitteilungen aus dem Gebiete des Seewesens* [Communications from the Field of Naval Affairs], Volume 41, in 1913, illustrates the careful considerations given to stability and lift during water takeoff and landing. These elements were essential to the operational success of seaplanes and significantly influenced Mickl's contributions to the development of Lohner seaplanes for the *k.u.k. Kriegsmarine*. His work on float design and hydrodynamic performance laid the foundation for improving the reliability and effectiveness of naval aviation.

ranks in the early 1900s mirrored the Austro-Hungarian Empire's rich ethnic diversity, reflecting the empire's complex societal structure. This diverse group included Germans, Hungarians, Croats, Serbs, Italians, Czechs, Slovenes, Poles, and Romanians. To ensure clear communication across this multicultural force, German was standardized as the command language, crucial for maintaining operational efficiency and safety aboard ships. Rigorous training and regular drills reinforced discipline and enabled all crew members to respond effectively to commands, ensuring the safety and success of naval operations. Within this diverse amalgamation of cultures, the innovative domain of naval aviation began to take shape in the *k.u.k. Kriegsmarine*.[29]

Operational Integration and Readiness

As the Austro-Hungarian Empire ventured into the 20th century, the operational imperatives of its maritime frontiers propelled the *k.u.k. Kriegsmarine* to embrace naval aviation. This transition became essential despite initial setbacks, such as base dependency and unpredictable Adriatic weather. The region posed unique meteorological challenges that invariably impacted naval operations. Two notable weather phenomena, the Bora and the Sirocco winds, frequently tested early aviators' capabilities. The Bora, characterized by cold, gusty winds from the Dinaric Alps, and the Sirocco, bringing warm, humid air from the Sahara, often created challenging flying conditions, demanding a high level of skill and adaptability from pilots.[30]

To overcome these challenges, the *k.u.k. Kriegsmarine*, under leaders like *Admiral* Rudolf *Graf* Montecuccoli, implemented measures to enhance aeroplane performance and reliability. The Pola Naval Arsenal played a crucial role by conducting extensive tests and modifications on their seaplanes.

Aeroplane structures were reinforced with additional bracing and more durable materials to endure sudden gusts and turbulent air. Engineers concentrated on improving engine performance and reliability, including experimenting with different engine models to ensure consistent power output. Enhanced cooling systems were developed to prevent overheating during extended operations, a critical step in addressing the challenges of early seaplane development. Notable advancements were achieved in aeroplanes like the

Above: Gottfried *Freiherr* von Banfield (left) and Franz Mikuleczky (right) were pivotal figures in Austro-Hungarian naval aviation. By the summer of 1914, as the *k.u.k. Seeflugwesen* solidified its role within the military, officers like Mikuleczky ensured the seamless transition of command and operational readiness. Banfield, along with Mikuleczky and other officers, played a critical role in acquiring new seaplanes from foreign manufacturers, including testing and selecting Donnet-Lévêque and Paulhan-Curtiss flying boats from France.

Lohner, and various floatplane models tested in Pola, which received critical structural and engine modifications. These improvements were essential for effective reconnaissance and patrol missions, ensuring reliable operation even in the Adriatic's unpredictable weather.[31]

Successfully adapting to the Adriatic's variable weather conditions reflected the navy's broader commitment to leveraging aviation technology for military control. The use of aeroplanes allowed the navy to extend its reach and maintain surveillance over critical maritime zones, ensuring the security of their coastal territories and supporting broader military objectives.

In December 1912, the *k.u.k. Seeflugwesen* undertook extensive efforts to purchase additional flying boats from foreign manufacturers, reflecting an expansion in both equipment and training programs. The purchase of flying equipment was entrusted to *Lschlt* Wenzel Leopold Woseček, who, along with *Frglt* Gottfried *Freiherr* von Banfield and *Lschlt* Viktor Klobučar Rukavina von Bunič, traveled to France to test and acquire new seaplane models. Klobučar, having previously visited France from 29 December 1911 to 15 January 1912, to inspect various flight facilities and later participating in key test flights and inspections in 1912, was deeply involved in the selection process. This effort included the acquisition of two Donnet-Lévêque flying boats, designated *S.8* and *S.10*, which were taken possession of by Woseček and Banfield in Paris on 2 December 1912, and another two Donnet-Lévêque flying boats, designated *S.11 and S.12*, which were acquired on 21 December 1912. A Paulhan-Curtiss flying boat, designated *S.9*, was acquired on 1 January 1913. On 7 January 1913, Klobučar visited Voisin in Paris to examine a floatplane. In March 1913, the Voisin floatplane was ordered and was designated *S.14*. At the same time, a D'Artois flying

Above: Paulhan-Curtiss flying boat *S.9*, acquired by the *k.u.k. Seeflugwesen* on 1 January 1913, was part of a broader initiative to modernize Austro-Hungarian naval aviation. Along with models like the Donnet-Lévêque, this seaplane was carefully selected and tested by officers including *Lschlt* Wenzel Leopold Woseč ek, *Frglt* Gottfried *Freiherr* von Banfield, and *Lschlt* Viktor Klobučar. Its acquisition represented a crucial step toward expanding the navy's access to aeroplanes for training and operational versatility.

boat, designated *S.13*, and another Paulhan-Curtiss flying boat, designated *S.15*, were ordered from France. These aeroplanes were tested extensively in Pola in early 1913.[32]

By 1913, the *k.u.k. Seeflugwesen* had ambitious plans to purchase 46 seaplanes, which were to be distributed as follows: 16 for the *Kriegsmarine*'s flagship, rapid cruisers, and armored cruisers, 16 for Pola and other coastal flight stations, 10 in reserve, and four large reconnaissance aeroplanes. However, this planned expansion did not fully materialize, primarily due to the preference for placing orders with the domestic economy, which faced challenges in meeting the demands. The appearance of foreign-built flying boats in the *k.u.k. Kriegsmarine* emphasized the need for improved domestic production capabilities. This, in turn, prompted investments in local aviation infrastructure and expertise, with factories like the Motoren-Luftfahrzeug-Gesellschaft mbH (MLG) increasing their capacity to produce aeroplanes tailored to the *k.u.k. Kriegsmarine* requirements.[33]

Seaplane production commenced at the Naval ArsenalPola workshops, which became a critical hub for the construction and maintenance of naval aeroplanes. This facility enabled the Navy to build and repair its seaplane fleet more efficiently, ensuring that the growing number of seaplanes could be adequately supported. In April 1913, the MLG factory received an order for flying boats based on Josef Mickl's designs. These flying boats, including the Lohner *E*, *M*, and *Mn* models, became vital to the *k.u.k. Seeflugwesen*, serving as essential "workhorses" during World War I. To accommodate the expansion, a new training station was established on Kosada Island in the Pola area, addressing the overcrowding at St. Caterina.[34]

Realizing the difficulty of planes taking off from naval vessels, the *k.u.k. Kriegsmarine* decided to focus on water launches. The command looked carefully at the two types of seaplanes: floatplanes and flying boats. After thorough studies and comparisons with seaplanes used by foreign powers, the *k.u.k. Kriegsmarine* chose to develop flying boats, better suited for the maritime environments they frequently encountered, as opposed to aeroplanes with wheeled chassis and floats. To implement these plans, establishing test aeroplanes, bases, and workshops was essential, alongside the training of officers and naval personnel.

During the First Balkan War, an international naval blockade of the Kingdom of Montenegro's coast took place from October 1912 to May 1913.[35] The *k.u.k. Kriegsmarine* was part of this contingent, and along with ships, seaplanes were assigned to the mission, performing reconnaissance and mine detection, marking the first use of naval aviation in active military

Left: *S.8* Donnet-Lévêque flying boat deployed to Montenegro, showing the hull being transported on a small craft, which highlights the logistical efforts required to move seaplanes during the Montenegro deployment. The *k.u.k. Seeflugwesen* assigned the *S.8*, along with other flying boats, to the mission at Teodo in 1913 during the First Balkan War. The seaplanes performed critical reconnaissance and mine detection operations, marking the first use of Austro-Hungarian naval aviation in active military operations.

operations. A temporary station was established in Teodo, Montenegro, within the strategic and operational facilities in the Bay of Kotor, active from 23 March to 28 May 1913. It was commanded by *Lschlt* Wenzel Leopold Woseček, with pilot officers *Lschlt* Hugo Ockermüller and *Frglt* Heinrich Huss flying the three Donnet-Lévêque flying boats (*S.8*, *S.10*, and *S.12*). The three flying boats conducted operations, performing visual observations, taking photographs, and delivering messages along the coast. Their success demonstrated the value of naval aviation. Originally assigned to the *2. Schlachtschiffdivision* [2nd Battleship Squadron] of the *k.u.k. Kriegsmarine* in Pola on 24 April 1913, they were transferred to Teodo on 26 April 1913. The crew performed 64 reconnaissance flights. This aviation detachment returned to Pola on 30 May 1913.[36]

By the end of 1914, the *k.u.k. Seeflugwesen* had trained seven NCO naval pilots, 11 officer pilots, and 150 flying and technical personnel, conducting training flights over a designated zone in the Adriatic. The increasing range of naval seaplanes, approximately 100 km, allowed them to undertake extensive military exercises. For instance, in January 1913, a group flight covering 240 km took place with three Donnet-Lévêque flying boats flying from Pola to Trieste and back. On 21 February 1913, a successful mine detection exercise was conducted using two Donnet-Lévêque flying boats (*S.10* and *S.11*) in simulated war conditions. Another mine search exercise took place on 12 March 1913, involving two seaplanes covering the route from Pola to Brioni and von Fasana Channel and back. Throughout the year, seven aviation officers completed 621 flights, with 13 to 18 flights covering

Above: *S.10* Donnet-Lévêque flying boat pictured on a rampway in Montenegro, positioned nose out, ready for a water departure. Like the *S.8*, the *S.10* played a key role during the 1913 naval blockade, performing reconnaissance flights from the base in Teodo. The use of these seaplanes demonstrated the growing capabilities of naval aviation, with *S.10* conducting numerous flights as part of the Austro-Hungarian naval aviation's contribution to the operation.

longer distances over the Adriatic.[37]

In June 1913, the first night flight in the history of the *k.u.k. Seeflugwesen* was performed by the Donnet-Lévêque *S.12* flying boat. In November 1913, onboard wireless tests were conducted on the Voisin *S.14* floatplane. By early 1914, the *k.u.k. Seeflugwesen* continued expanding its capabilities with the construction of new aviation facilities and the acquisition of flying boats, such as the Lohner *E* and *M* series. These flying boats played crucial roles in reconnaissance and patrol missions, demonstrating the importance of naval aviation in the Austro-Hungarian military framework.[38]

In 1913, Heinrich Huss, now promoted to *Lschlt*, began flying *S.10*, a Donnet-Lévêque seaplane at the Pola air station until the outbreak of the First World War. The Donnet-Lévêque flying boats, such as the machines with production numbers *S.8*, *S.10*, *S.11*, and *S.12*, were equipped with type *A*s a 50-hp Gnome seven-cylinder engine and *C*s an 80-hp Gnome seven-cylinder and were primarily used for training and reconnaissance missions. Huss made more than 68 flights in Pola and Teodo, with flight durations ranging from 30–40 minutes to some extending up to two hours. These seaplanes, ordered in late 1912 and delivered in early 1913, remained in service at the time of hostilities in 1914.[39]

By June 1914, night flights and exercises involving searchlights were conducted to enhance operational readiness. The training centers at Experimental Flight Station St. Caterina and Training Flight Station Kosada were highly active, preparing personnel for a variety of military scenarios. The list of flying accidents during this period, which included notable incidents with Donnet-Lévêque flying boats, underscored the challenges faced in developing and operating early naval aeroplanes. In 1914 alone, there were at least 14 separate accidents involving naval aeroplanes. These incidents, ranging from pilot errors to mechanical failures, reflected the inherent risks and technical difficulties of early aviation, with frequent mishaps indicative of the trial-and-error nature of early flight training and the ongoing development of aeroplane technology.[40]

By the summer of 1914, as the political situation in Europe deteriorated, the *k.u.k. Seeflugwesen* had solidified its position as an element of the Austro-Hungarian military. The naval aviation units, equipped with a range of seaplanes and supported by trained personnel, were poised to play a role in the upcoming conflict. The transition of command within the *k.u.k. Seeflugwesen*, involving officers like *Lschlt* Franz

Above: Voisin L *S.14* floatplane, acquired by the *k.u.k. Seeflugwesen*, was part of the navy's early efforts to enhance its reconnaissance and communication capabilities. In November 1913, *S.14* was used to conduct onboard wireless tests, marking an important milestone in naval aviation's evolution. These experiments, along with the acquisition of additional machines like the Lohner E and M series, reflected the continued expansion of the Austro-Hungarian naval aviation program.

Mikuleczky and *Lschlt* Woseček, ensured a seamless continuation of leadership and maintained operational readiness. These officers played critical roles during their respective tenures, guiding the naval air service through its formative years and preparing it to meet the evolving challenges of naval warfare. Both Mikuleczky's command and Woseček's subsequent leadership exemplified the deliberate foresight and adaptability required to advance the *k.u.k. Seeflugwesen's* capabilities during the early wartime period.[41]

The rapid expansion and development of the *k.u.k. Seeflugwesen* by 1914 exemplified the shift towards incorporating aviation technology into naval operations. This transformative growth, driven by visionary leaders and enabled by the necessary financial resources for training and equipment, marked an important chapter in the history of Austro-Hungarian naval aviation.

Endnotes

1. The Battle of Lissa, fought on 20 July 1866, was a crucial engagement between the Austrian and Italian navies during the Third Italian War of Independence. Commanded by *Konteradmiral* Wilhelm von Tegetthoff, the Austrian fleet secured a surprising victory against a numerically superior Italian force. This battle underscored the importance of innovative naval tactics and the effective use of ironclad warships. The outcome demonstrated that technological advancements and naval acumen were essential for maritime parity. This victory not only reinforced the Austrian Empire's commitment to naval modernization but also highlighted the necessity of naval capabilities to protect and expand its maritime interests.

Above: Heinrich Huss sits on Donnet-Lévêque *S.10* in Teodo, 1913, with a second flying boat visible behind him. The nose-in position of the seaplane suggests its return from a mission. Huss, recently promoted to *Lschlt*, flew *S.10* regularly at both Pola and Teodo, performing training and reconnaissance flights. The Donnet-Lévêque flying boats, including *S.8, S.10, S.11*, and *S.12*, were equipped with either a 50 hp or 80 hp Gnôme engine. Huss logged over 68 flights, with durations ranging from 30 minutes to up to 2 hours. These aeroplanes remained in active service when World War I began.

2. Cowburn, P. M., "Late 19th-Century Naval Journals and Memoirs as History," *Naval Historical Society of Australia*, August 1972
3. *SMS* or *S.M.S.*" stands for "*Seiner Majestät Schiff*," meaning "His Majesty's Ship." This prefix was used for ships of the *k.u.k. Kriegsmarine* (Austro-Hungarian Navy) to denote royal and imperial vessels under the Austro-Hungarian Empire.
4. SAW-KA, W. Nemetz, *Die Militärluftfahrt Oesterreich Ungarns im Jahre* 1914; *k.u.k. Kriegsmarine* Archives, "Observation Balloon Tests and Exercises 1902-1907": Detailed records of the initial balloon tests and exercises conducted by the navy.
5. Peter Schupita, *Die k. u. k. Seeflieger: Chronik und Dokumentation der österreichisch-ungarischen Marineluftwaffe 1911-1918*, 9.
6. *Lschlt* Wenzel Leopold Woseček (Václav Leopold Voseček), an Austro-Hungarian Naval officer, served notably in naval aviation. Known by his Czech name, Václav Leopold Voseček, and his German name, Wenzel Leopold Woseček, he was born on 26 September 1882 in Hořice, Bohemia, Austria-Hungary. He played a significant role as a pilot and commander in early naval aviation efforts: Dieter Winkler, "Ing. Heinrich Huss, Marineflieger und Korvettenkapitän." *Flagge* [The flag], ⊠. 3/4, 2010, pp. 29–30.
7. *Admiral* Rudolf *Graf* Montecuccoli, "Naval Aviation Reports, 1910-1914": Reports and directives issued by *Admiral* Montecuccoli on the development and implementation of naval aviation strategies.
8. Valuation of Austro-Hungarian kronen to USD (1914): 1 kronen ≈ 0.203 USD : 100,000 kronen ≈ 20,300 USD
9. Valuation of Austro-Hungarian kronen to USD (1914): 1 kronen ≈ 0.203 USD.: 312.4 million kronen ≈ 63.4 million USD.
10. *Lschlt* Viktor Klobučar Rukavina von Bunić. In 1909, he underwent civilian pilot training in France, funding the entire endeavor himself : Charles W. Koburger, Jr., *The Central Powers in the Adriatic, 1914–1918: War in a Narrow Sea*, Westport, CT: Praeger Publishers, 18: Leonard Eleršek, *Homo volans: rani hrvatski avijatičari 1554. - 1927: ilustrirana kronika hrvatskog zrakoplovstva sa zbirkom fotografija Josipa Novaka*, Selbstverl., 2009. 119–120.
11. Peter Schupita, 11.
12. Heinrich Huß (Huss) was born on 13 March 1881 in Pilsen, Bohemia. He had a distinguished career in the Austro-Hungarian Navy. Huß passed away on

Above: A Donnet-Lévêque flying boat under tow by SMS *Lika*, a *Tátra*-class destroyer, in 1914. While the exact reason for the tow is not documented, it is possible that an engine issue or other mechanical failure required transport back to a station for repair, not an uncommon occurrence in seaplane operations. This scene reflects the logistical challenges faced by Austro-Hungarian naval aviation as it integrated new technologies and sought to maintain operational effectiveness during its formative years.

21 June 1966, having contributed significantly to his nation's naval history during and after World War I. After becoming one of the early pilots in the *k.u.k. Kriegsmarine*, Huß continued to serve with distinction in various capacities. Following his involvement in naval aviation, he transitioned back to more traditional naval roles, eventually achieving the rank of *Korvettenkapitän* [Corvette Captain] and holding several important positions within the Austro-Hungarian Navy, particularly in the areas of training and command. At the outset of World War I in 1914, *Lschlt* Heinrich Huß took command of Torpedo Boat (TB) *5* and led a torpedo boat group comprising TB *18* and TB *19*. By February 1916, he was commanding TB *58 T*, and from November 1916 until the end of the war, he commanded TB *97 F.*

13. Peter Schupita, 11: Jiří Rajlich, *Příslušníci rakousko-uherského letectva z českých zemí* [Members of the Austro-Hungarian air force from the Czech lands]. Dizertační práce [Dissertation], vedoucí [supervisor] Županič, Jan. Praha: Univerzita Karlova, Filozofická fakulta, Ústav světových dějin, 2018, 54–55.
14. Peter Schupita, 18.
15. Peter Schupita, 18.
16. Reinhard Keimel, *Flugzeuge der österreichischen Firma Lohner, 1909-1923* [Planes of the Austrian company Lohner], Sonderheft der Blätter für Technikgeschichte, 1990, 240–243.
17. Peter Schupita, 18.
18. Marcus Zelezny, *Aviation in Wiener Neustadt (1909–1912)*, Stanger Verlag, 2022, 214: Reinhard Keimel, *Österreichs Luftfahrzeuge: Geschichte der Luftfahrt von den Anfängen bis Ende 1918*, [Austria's aeroplane: the history of aviation from the beginning to the end of 1918] H. Weishaupt Verlag, 1981, 403: Jerzy B. Cynk, *Polish Aircraft, 1893-1939*. London: Putnam, 1971, 67.
19. Reinhard Keimel, 243–249.
20. Josef Mickl, "Das Schwimmkufenproblem" [The issue of float design], Mitteilungen aus dem Gebiete des Seewesens, vol. 41 (Pola: k. u. k. Marinetechnischen Komitee, Marinebibliothek, 1913), 58, 268: Mickl's first seaplane was constructed in the Naval Arsenal. Subsequently, Mickl, in collaboration with Igo Etrich

and Karl Paulal, played a pivotal role in the design of the highly successful Lohner seaplanes and led the Konstruktionsbüro des Seefliegerkorps [Seaplilot Corps design office].

21. James J. Davilla and Arthur M. Soltan, *French Aircraft of the First World War*, CA, Flying Machines Press, 1997, 190–191: Jiří Rajlich, 56.
22. Peter Schupita, 11.
23. Peter Schupita, 11.
24. Dieter Winkler, 29-30 : Hans Hugo Sokol and Theodor Braun, *Österreich-Ungarns Seekrieg 1914-18* [The Austro-Hungarian sea war 1914–1918], Akademische Druck-und Verlagsanstalt, 1967, 2 volumes.
25. Valuation of Austro-Hungarian kronen to USD (1914): 1 kronen ≈ 0.203 USD : 170 million kronen ≈ 34.5 million USD : 205 million kronen ≈ 41.6 million USD : 484,000 kronen ≈ 98,252 USD.
26. Peter Schupita, 24.
27. Valuation of Austro-Hungarian kronen to USD (1914): 1 kronen ≈ 0.203 USD : 3 million kronen ≈ 609,000 USD : 134 million kronen ≈ 27.2 million USD : Milan N. Vego, *Austro-Hungarian Naval Policy, 1904–14*, London: Frank Cass, 1996, 136–141, 161–163.
28. Jiří Rajlich, 58-59.
29. *Beyond Nationalism. A History of the Habsburg Officers Corps, 1848-1918*, New York-Oxford 1992, 201: Milan N. Vego, 25–26.
30. The Bora is a cold, dry wind that descends from the northeast across the Adriatic Sea, originating in the Dinaric Alps. It is known for its strong gusts, which can reach speeds of up to 220 km/h (136 mph). The Bora typically occurs during the winter months but can happen at any time of the year. Its strength and sudden onset pose significant challenges to both maritime and aviation operations in the region. The Sirocco, on the other hand, is a warm, humid wind that originates in the Sahara Desert and travels northward across the Mediterranean Sea. This wind often brings sand and dust, reducing visibility and creating challenging flying conditions. The Sirocco can occur throughout the year but is most common in the spring and fall. Both winds are episodic and highly variable, making them difficult to predict and manage during naval and aviation operations.
31. Peter Schupita, 14.
32. James J. Davilla and Arthur M. Soltan, *French Aircraft of the First World War*: Jiří Rajlich, 56.
33. Peter Schupita, 14 : Reinhard Keimel, 249–255.
34. Peter Schupita, 26–28.

Above: *k.u.k. Seeflugwesen "Fliegerdienst"* [Aviator Service] *"Spezialitätenabzeichen"* [Specialty badge], introduced 1913. This badge featured the winged propeller, combined with the crown and wave motifs, reflecting the integration of naval aviation into the *k.u.k. Kriegsmarine*. For the ranks of *Fliegerquartiermeister* [Flight Quartermaster] and *Fliegermaat* [Flight Mate], the badge was made of yellow and dark blue sheep wool and sewn onto the middle of the left upper sleeve. For higher non-commissioned officer ranks, including *Flieguntermeister* [Assistant Flight Master], *Fliegermeister* [Flight Master], and *Oberstabsfliegermeister* [Senior Staff Flight Master], the same design was embroidered in gilt silver wire.

35. Milan N. Vego, 149–152.
36. Peter Schupita, 14.
37. E. R. Hooton, *Prelude to the First World War: The Balkan Wars 1912–1913*, Fonthill Media, 2014.
38. Peter Schupita, 14: Reinhard Keimel, 252.
39. Dieter Winkler, 29-30: James J. Davilla and Arthur M. Soltan, 190–191.
40. Peter Schupita, 99.
41. Peter Schupita, 14, 16, 139.

9. Fischamend's Technological Innovations

Above: Wireless telegraphy equipment installed in the Etrich Taube series "F," named 'ALLONS,' developed by Prof. Kann. Following the 1913 *Das Offiziersfliegen* [Officers' Flying] event, extensive trials began at Fischamend to improve artillery coordination using wireless telegraphy. Under the leadership of *Hptm* Ferdinand *Ritter* Cavallar von Grabensprung, the first operational wireless units were successfully tested for directing artillery fire, with the equipment achieving a range of 300 km. *Credit: Reinhard Masajdek*

The Fischamend *Versuchsabteilung* [Experimental Detachment] of the Aeronautics Department, under the command of *Hptm* Ferdinand *Ritter* Cavallar von Grabensprung since 1913, aimed to transform aeroplanes into weapons. This was undertaken with a number of experiments involving various aeroplanes.

They were equipped with machine guns and had devices installed to permit bomb-dropping, as well as a variety of munitions from small grenades to larger ordnance utilizing a mount with an internal release mechanism. In addition, variety of aiming devices, from simple to complex, were tested to adjust for speed angle and drift. Although aerial bombing trials conducted produced mixed results, further testing and devices for accurate targeting was already planned. The mounting of a machine gun on aeroplanes such as the two-seated Lohner *Pfeilflieger B* and *C* showed immediate promise. Initially, several lighter guns, including the Lewis designed by the American Colonel Isaac Newton Lewis and produced in Belgium, were considered. However, due to higher fire rate and ready availability, the Schwarzlose M.07/12 machine gun was preferred. It was modified for aerial use, which reduced its weight to 25.6 kg, including 250 rounds of ammunition.[1]

In the Austro-Hungarian Empire, it was acknowledged that the most effective means of communication between aeroplanes and ground forces was through wireless telegraphy. A significant area

Above: The Scheimpflug-Kammerer *Panoramaapparat* [Panoramic Apparatus], an innovative aerial photographic device developed by Theodor Scheimpflug and *Ingenieur* Gustav Kammerer, underwent evaluation at Fischamend in 1913 for reconnaissance purposes. Designed to advance aerial photography, the device was subjected to a series of trials using airships and balloons. These experiments culminated in 1914, when the apparatus was deployed during the ill-fated test flight of the airship M.*III*.

of research at the Fischamend focused on improving aeroplane observation for artillery fire adjustment, which necessitated direct communication between aeroplanes and ground forces. With the help of Professor Kann, experiments with wireless units began in 1913, leading to the development and refinement of the first experimental devices specifically designed for aeroplanes.[2] Throughout the spring and fall of 1913, a series of tests were conducted with various aeroplanes and wireless devices. Despite the weight of the equipment, the devices demonstrated the potential for effective communication during reconnaissance missions. Numerous reports were submitted to the *k.u.k. Kriegsministerium* detailing the results of these experiments. To make the devices suitable for wartime use, it was determined that the equipment

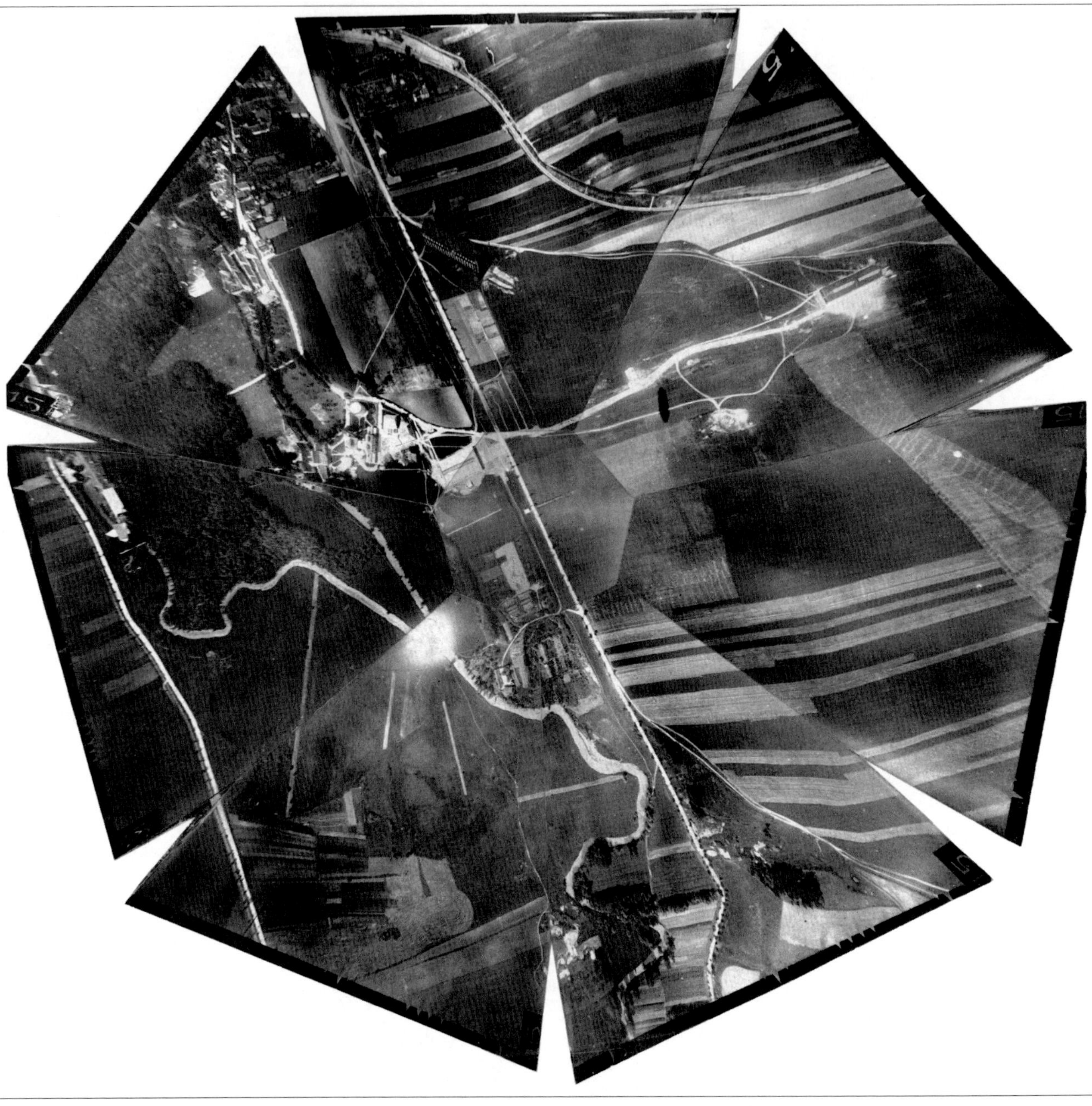

Above: An aerial photograph created by the Scheimpflug-Kammerer system, composed of seven images taken simultaneously to provide a panoramic view. The camera utilized an innovative system of multiple lenses arranged in a circular configuration, each capturing a segment of the surrounding terrain at the same moment. This synchronized method minimized distortion while maximizing coverage. In the 2 o'clock position, the shadow of the dirigible from which the photograph was taken is visible. The image highlights the advanced aerophotography techniques being pioneered in Austria-Hungary during this period. *Credit: Rudolf Ster*

needed to be lighter and standardized procedures for fire adjustment had to be established using a limited number of predefined signals. Since messages would be transmitted via Morse code, it was essential for both observers and ground artillery officers to have adequate understanding and training.

In March 1914, the first field-usable flight wireless unit, designed by Professor Kann and constructed by Siemens & Halske in Vienna, became available. In May 1914, the device was installed in the Lohner-Etrich *Taube* by *Hptm* von Cavallar, the head of the testing department at Fischamend.[3] This series of successful tests led to the first practical fire control tests for military air-to-ground communication in Austria-Hungary. Von Cavallar was able to direct the fire of a 15 cm M 80 cannon and a 30.5 cm M 11 mortar

Right: The ICA Minimum Palmos 6x9 cm camera, equipped with a Tessar lens, was tested in 1913 at the Fischamend laboratory as part of pioneering efforts to adapt commercial photographic equipment for aerial reconnaissance. Known for its portability and compact design, the camera was considered suitable for mounting in aeroplane or being operated by hand during flight. However, the limitations of its resolution and field of view in capturing the detailed imagery required for tactical and strategic purposes became evident during evaluations. These trials highlighted the need for specialized aerial cameras, ultimately leading to the development of more robust, purpose-built equipment for reconnaissance missions.

battery using wireless communication. The onboard transmitter weighed 35 kg and had a range of 300 km.[4] These successful tests contributed to the establishment of the first ground radio station at Fischamend, specifically designed for aeroplane communications.[5]

The assessment of aerial photography within military aviation presented a unique case study, with notable disparities between German and Austro-Hungarian approaches. At the German military airfield in Döberitz, Germany, *Ltn* Carl Fink, a German officer, independently experimented with aerial photography without any directive or support from his superiors. In contrast, the Austro-Hungarian Aeronautics Department command at Fischamend conducted a comprehensive test program employing the Scheimpflug-Kammerer system and other devices. Official training for pilots and observers commenced following the first successful photographs taken during 1912 maneuvers.

At Fischamend, an aerophotography laboratory had been established, building on prior experiments with balloon and airship photography using the Scheimpflug-Kammerer aerial photographic apparatus. The laboratory was led by an expert from the Militärgeographisches Institut [Military Geography Institute] and supported by a small team of civilian photographers who were enlisted. In the summer of 1913, the evaluation of aerial cameras for their usefulness and quality was initiated. This assessment included testing the Scheimpflug-Kammerer, the Zeiss Fliegerkammer, the Ernemann AG camera, and the ICA Palmos camera, as well as a design created by the k.u.k. Militärgeographisches Institut.[6] In spring 1914, aerial camera tests were conducted, culminating in a test flight for aerial photography with the ascent of airship *M.III* on 20 June 1914, which unfortunately resulted in the Fischamend Körting catastrophe.

With the onset of World War I in August 1914, commercial Zeiss and Goerz handheld cameras with short-focus objectives, telescopic bellows, and 9 x 12 cm plates were hastily procured to equip the *FliKs*. The favored lenses were Zeiss-Tessar objectives with 15 cm focal lengths and an aperture ratio of 1 to 4.5. Unbeknownst to the Austro-Hungarians, German *Ltn* Fink had already determined that bellows cameras were unsuitable for aeroplane usage due to wind compression

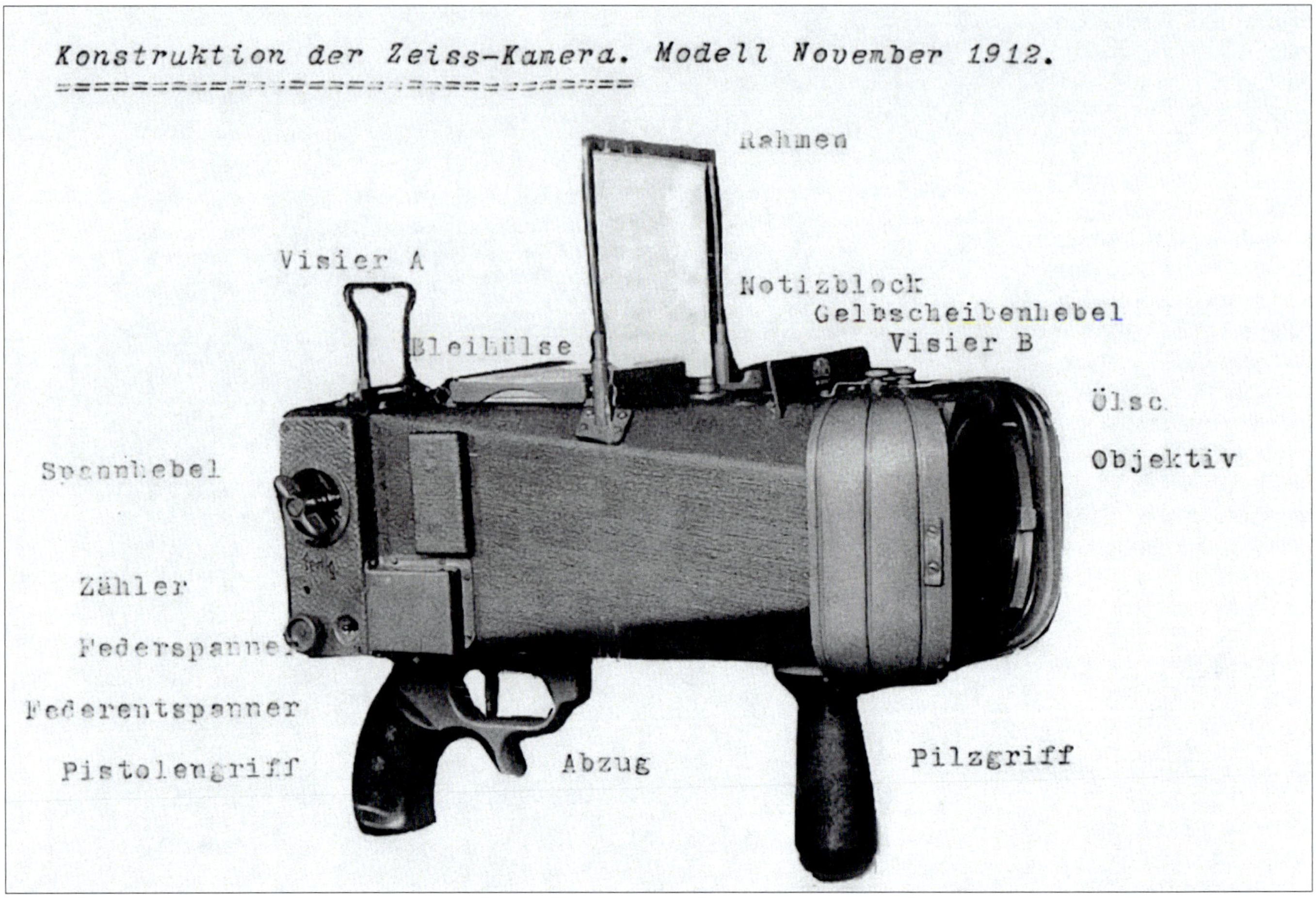

Above: The 1912 Zeiss *Fliegerkammer* [Aviation Camera], referred to as the *Revolver-Kamera* [Revolver Camera] by the Austro-Hungarians, was a variant of the German *Pistolenkammer* [Pistol Camera], initially devised by *Lt* Carl Fink, a member of the German *Fliegertruppe* [Flying Troops]. The German military had only limited quantities of these cameras, prompting Germany to impose an export ban at the outbreak of the war. This restriction forced the *k.u.k. Luftschifferabteilung* at Fischamend, specifically its testing division, to begin developing indigenous aerial photographic solutions. Although the Zeiss design was advanced for its time, more practical and durable models from German manufacturers like Goerz and Herzig were later preferred by the Austro-Hungarian military as aerial reconnaissance became increasingly critical. *Credit: Helmut Jäger.*

of the bellows and unreliable shutter mechanics. Furthermore, the short focal length and small 9 x 12 cm plates yielded images with insufficient tactical information.[7]

The German military possessed the suitable Pistolenkammer, devised by *Ltn* Carl Fink, albeit in limited quantities. Consequently, at the beginning of the war, Germany imposed an export ban on their Pistolenkammer in production. The *k.u.k.* War Ministry had to rely on domestically developed cameras, such as the sophisticated Scheimpflug-Kammerer, which was suitable for geodetics but ill-suited for harsh field conditions and therefore was never deployed. The Herzig company developed a stereo aerial camera with 15 or 18 cm focal lengths and 9 x 18 cm plates, which was selected and distributed to a few *FliK*s along with stereo viewing devices. However, these cameras were rarely utilized on missions, as higher-ranking officials showed little interest in photographs, preferring visual reports. By 1915, the Herzig cameras were replaced with German Zeiss and Goerz aerial cameras, which had been approved for export to Germany's ally, Austro-Hungary.[8]

Endnotes

1. Desoye, 68; Erwin Steinböck, *Bewaffnung und Ausrüstung der österreichisch-ungarischen Flugzeuge.* [Armament and equipment of Austro-Hungarian aircraft], Graz 1983, 7; George M. Chinn, *The Machine Gun History, Evolution, and Development of Manual, Automatic, and Airborne Repeating Weapons by Volume I*, Bureau of Ordnance Department of the Navy 1951, 275–299 (Lewis), 228–232 (Schwarzlose); Source: Interessengemeinschaft Luftfahrt Fischamend;

Above: The Goerz Anschutz 5×4 camera, also trialed in 1913, faced challenges in adapting to the unique demands of aerial photography. While initially employed in experimental flights, the camera's design, with its telescopic bellows and short-focus lens, proved inadequate for the precise reconnaissance work required by military operations, leading to the search for more suitable equipment.

correspondence with *Ing* Rudolf Ster; Archival materials on Luftfahrt Fischamend 1909–1913.

2. Source: Interessengemeinschaft Luftfahrt Fischamend; correspondence with *Ing* Rudolf Ster; Archival materials on Luftfahrt Fischamend 1909–1913.
3. Interessengemeinschaft Luftfahrt Fischamend; Rudolf Ster.
4. The 15 cm Kanone M 80 was a siege gun used by Austria-Hungary during World War I. Designed to replace the M 61 series of siege guns, the M 80 family of siege guns offered greater range and armor penetration than the older guns. The Škoda 30.5 cm Mörser M.11 is a siege howitzer produced by Škoda Works and used by the Austro-Hungarian Army during World War I and by Nazi Germany in World War II.
5. Steinböck, *Bewaffnung,*47–49; In December 1914, the first wireless units were delivered to *FliK*s *(Fliegerkompanien)* №.1, 10, and 15. These devices weighed approximately 50 kg and had a range of 300 km. During the winter of 1914/15, Professor Kann developed the G1 aeroplane transmitter, which weighed 42 kg, and the smaller g1 onboard transmitter, which weighed 18 kg and had a range of 80 km. Series production of these devices began in February 1915 at the Siemens & Halske factory in Vienna, with a production rate of 10–15 units per week.
6. Interessengemeinschaft Luftfahrt Fischamend; Rudolf Ster.
7. Jäger, *Erkundung*, 79.
8. Source: KA-Wien, *Das Lichtbildwesen der öst. - ung. Luftfahrtruppen 1914/18* [Photography in the Austro-Hungarian air service 1914/18]. Typographic manuscript from Kriegsarchiv Wien, 1.

10. Emergence of Aerodynamics in Austria-Hungary

Austria-Hungary, at the turn of the 20th century, was a nation marked by a burgeoning scientific and technological landscape. This era was characterized by a commitment to academic excellence and innovation, particularly within its leading institutions. These establishments played a crucial role in advancing the understanding of physics and engineering, setting the stage for significant contributions to the nascent field of aerodynamics. The scientific community in Austria-Hungary was driven by a blend of theoretical inquiry and practical application, fostering a culture of rigorous experimentation and groundbreaking discoveries.

In 1895, Friedrich *Ritter* von Lössl's publication, "The Laws of Air Resistance, Falling through the Air and Bird Flight," laid the groundwork for future aerodynamic research. By combining mathematical modeling with mechanical experiments, Lössl analyzed air resistance and flight dynamics. His principles were foundational and influenced subsequent pioneers, aiding in the design of early aeroplanes and airships.[1] Similarly, Wilhelm Kress's 1901 investigative effort into the power-to-weight ratio demonstrated a nuanced understanding of the principles of aerodynamics and early flight engineering. Aware that the engine for his *Drachenflieger* was both underpowered and overweight, Kress nonetheless proceeded with his attempt, showcasing his commitment to advancing the field. His detailed and accurate calculations were crucial in defining the requirements for powered flight, highlighting the critical importance of matching power output to aeroplane weight—a principle that remains fundamental in aviation design today.[2]

During this transformative period, universities and technical institutes evolved into centers of rigorous scientific discourse. The Technische Hochschule Wien [Vienna Polytechnic Institute], for instance, became a hub of innovation where classical physics merged with emerging engineering breakthroughs. This intellectual climate promoted a methodical approach to understanding natural phenomena, particularly the principles of flight and air movement.

In this environment, the theoretical foundations of aerodynamics began to solidify. The construction of a state-of-the-art wind tunnel in Vienna was instrumental in advancing the understanding of air

Above: Richard Knoller, the pioneering Austrian aerodynamicist whose work on the Knoller-Betz effect profoundly influenced early aerodynamics. His tenure at the Technische Hochschule Wien [Vienna Polytechnic Institute] marked a critical period in the scientific study of aviation within Austria-Hungary, bridging the gap between theory and practical application in aeroplane design.

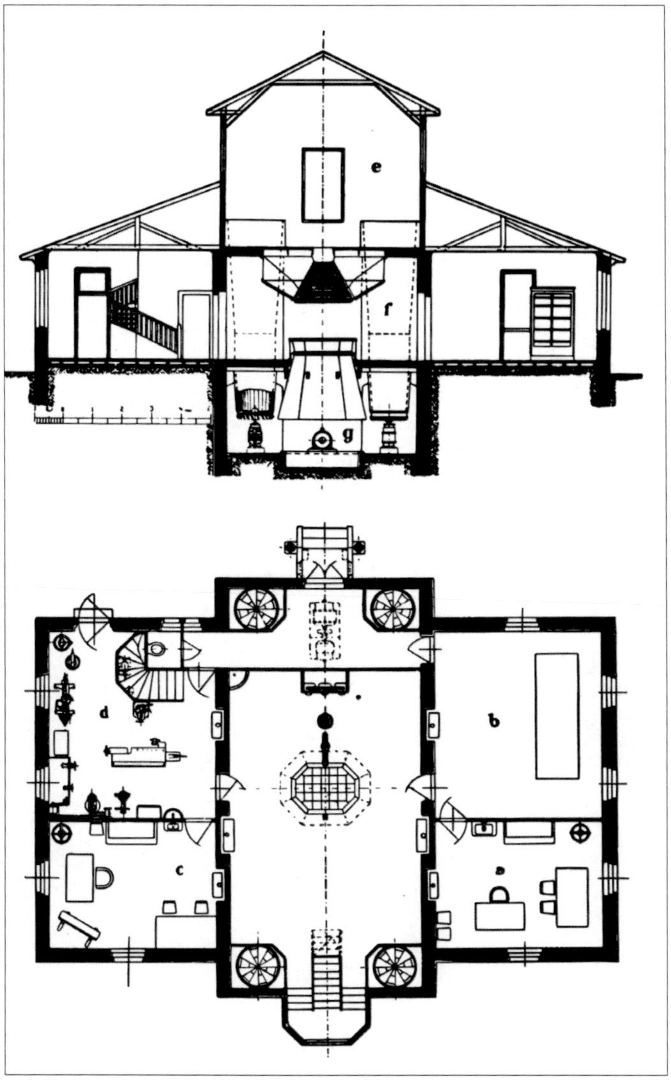

Above: The wind tunnel facility at the Vienna Polytechnic Institute, designed by Richard Knoller. This Eiffel-type open-circuit wind tunnel played a crucial role in advancing aerodynamic research and was essential for testing aeroplane designs and aerodynamic principles in pre-war Austria-Hungary.

dynamics and movement. These advancements were driven by institutional research initiatives, which provided the necessary intellectual and material resources for systematic studies.

The legacy of Austria-Hungary's contributions to aerodynamics was another link in the chain encompassing the growing international diffusion of knowledge. The research not only enhanced the scientific understanding of flight but also laid the groundwork for future innovations in aeronautics. By bridging the gap between theoretical research and practical engineering, Austria-Hungary established itself as a significant contributor to the early

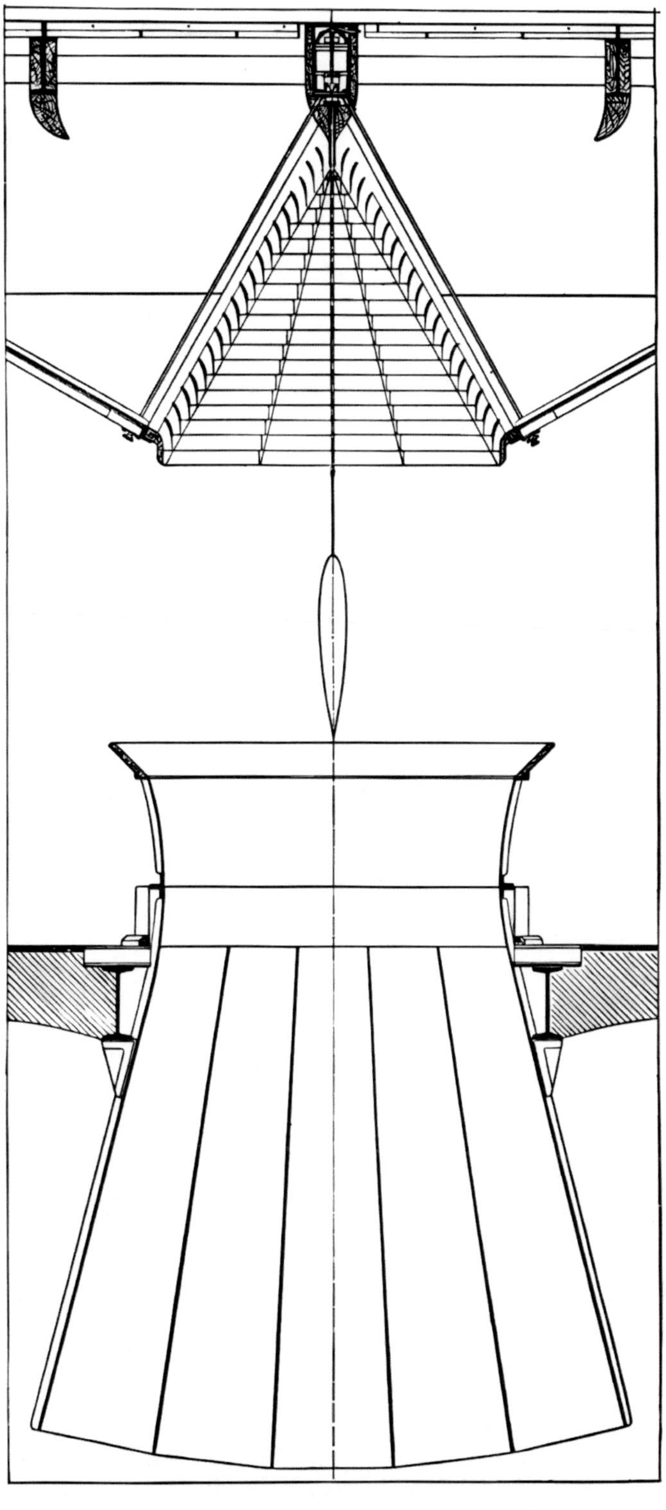

Above: Detailed sectional layout of the upper and lower sections of Knoller's wind tunnel at the Vienna Polytechnic Institute. This configuration allowed precise control of airflow and ensured accurate testing conditions for experiments in aerodynamics.

development of aerodynamics, influencing subsequent generations of scientists and engineers in the quest for mastering controlled flight.

Right: Instrumentation used within the wind tunnel for measuring aerodynamic forces. These devices provided critical data on drag, lift, and stability, enabling detailed analysis of aerodynamic properties and contributing to the refinement of aeroplane designs.

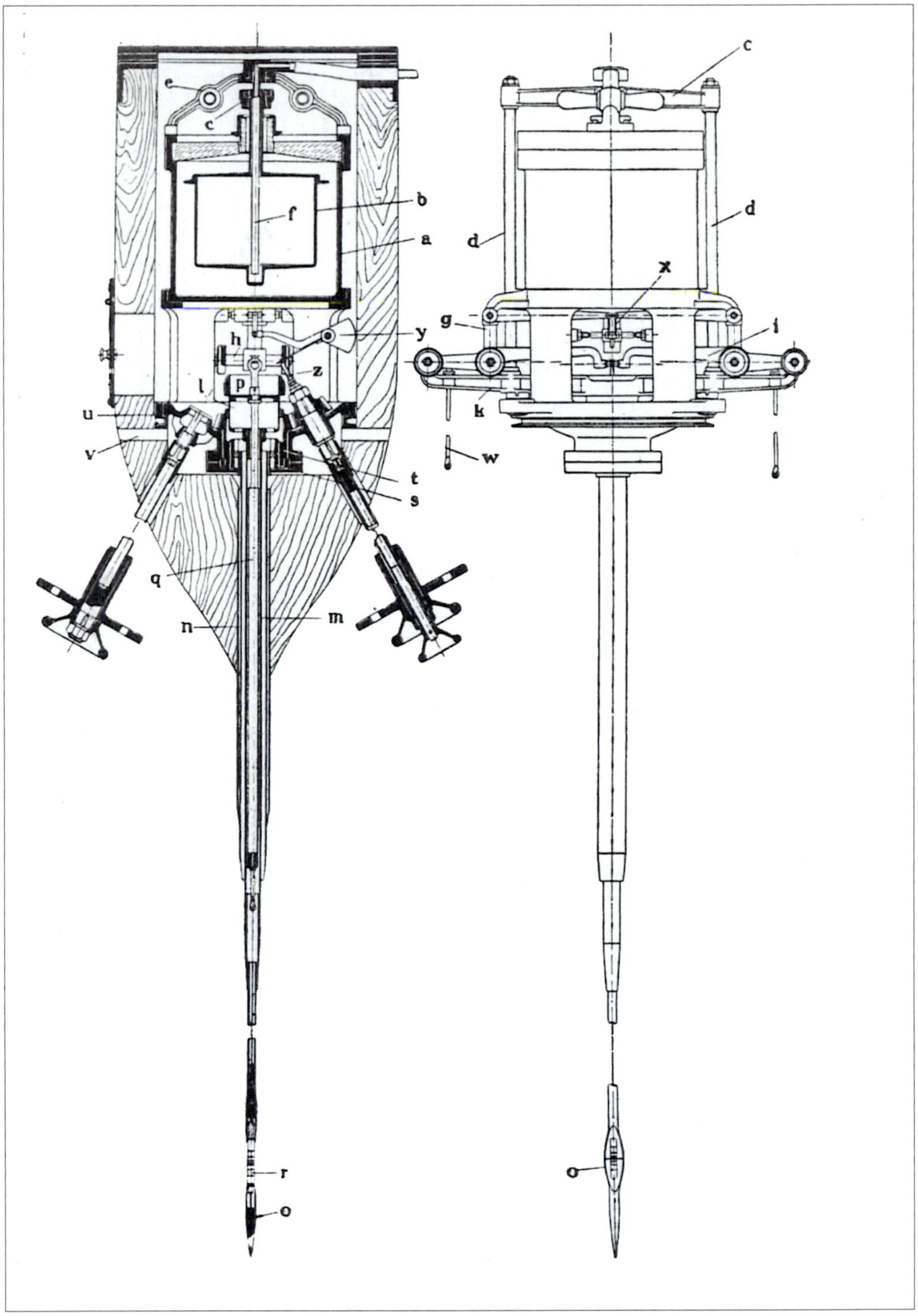

Richard Knoller: A Pioneering Figure in Aerodynamics

At the heart of Austria-Hungary's advancements in aerodynamics stood Richard Knoller, professor at the Vienna Polytechnic Institute. His profound understanding of air movement around wings and propellers led to the seminal discovery of the Knoller-Betz effect. This principle, which elucidated how rotating cylinders generate lift akin to the Magnus effect, became fundamental in the design of aeroplane wingtips and propellers, offering crucial insights for enhancing aeroplane performance and stability.

Richard Knoller's tenure beginning in 1909 as the first head of the newly established *Lehrkanzel für Luftschiffahrt und Automobilwesen* [Chair for Aeronautics and Automotive Engineering] at the

Above: Interior view of the wind tunnel testing room, showcasing the precision setup used for aerodynamic testing. The octagonal nozzle design and controlled airflow system exemplified Knoller's innovative approach to ensuring laminar flow and reducing turbulence during experiments.

Vienna Polytechnic Institute marked a pivotal era in the integration of academic rigor and practical engineering within the burgeoning technological landscape of early 20th century Austria-Hungary. His innovative and foundational approach to aerodynamics, deeply immersed in researching drag, propeller properties, and glider stability, mirrored a broader European shift toward more scientifically rigorous methodologies in aeroplane design and analysis. This work not only bridged academic inquiry with real-world engineering challenges but also positioned Knoller at the forefront of a significant transformation in how aerodynamic phenomena were understood and applied in the design of safer and more efficient aeroplanes.

The construction of the wind tunnel at the Vienna Polytechnic Institute between 1911 and 1914 was perhaps one of Knoller's most noteworthy contributions. This facility was notable for its Eiffel-type open-circuit design, involving a direct airstream setup rather than a closed loop, which minimized external influences on the air properties, such as temperature and pressure, that could affect experimental results. The vertical tunnel featured an innovative octagonal nozzle with adjustable guide vanes, designed by Knoller himself, to precisely control the airflow and reduce turbulence entering the test section. This was critical for ensuring laminar flow conditions, essential for accurate aerodynamic testing. The air entered this section as a free jet from a

Above: Balance and measurement devices used in Knoller's wind tunnel, essential for assessing drag and lift forces on aerodynamic models. These finely tuned instruments allowed for accurate data collection and were pivotal in understanding the aerodynamic behaviors of early aeroplanes.

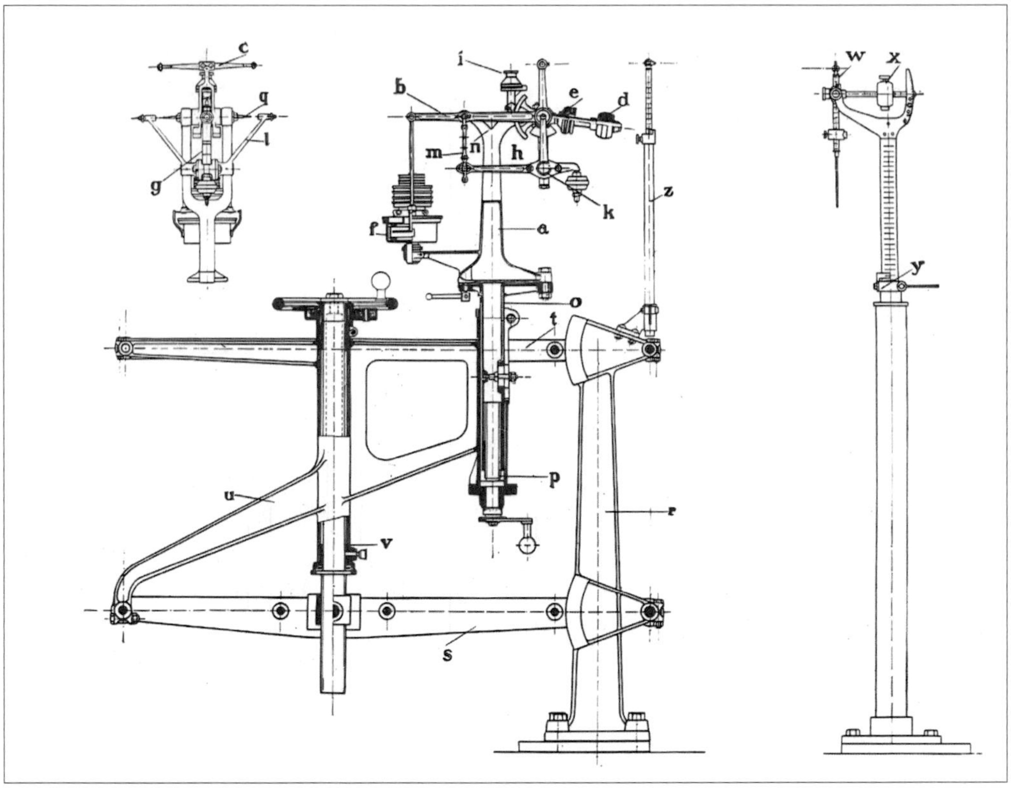

large calming room, ensuring smooth and controlled airflow. The tunnel's design facilitated easy access to the models being tested, streamlining the study of aerodynamic phenomena. Knoller's wind tunnel was not just a tool for routine testing; it was a hub of innovation. It allowed for experiments that not only refined the designs of aeroplanes but also contributed to a deeper understanding of fundamental aerodynamic principles. The data gathered in this tunnel helped to validate theories about airfoil performance, drag coefficients, and stability characteristics of various aeroplane configurations. Moreover, the wind tunnel's construction reflected the technological ambitions of the era. It was built with robust materials and featured advanced electrical systems for controlling the tunnel's operational parameters, such as wind speed and air temperature. The integration of these systems demonstrated a forward-thinking approach to experimental design, blending mechanical engineering with emerging electrical technologies to create a state-of-the-art research facility.

Knoller's meticulous attention to detail extended to the measurement techniques employed within the tunnel. A mercury float balance measured drag, acting vertically, while lift, acting horizontally, was measured using a mechanical scale that allowed for adjustments in the angle of attack and precise determinations of the center of pressure. These innovative measurement methods provided highly accurate data crucial for understanding the physical behaviors affecting aeroplane stability and efficiency.

Knoller's investigations, especially into the aerodynamics of gliders, provided critical insights that later influenced the design and development of safer and more reliable aeroplanes. His theoretical work on propeller design also contributed to the development of more efficient propulsion systems, which became essential as aeroplanes required more power to carry heavier loads and achieve greater speeds.

The legacy of Knoller's early work transcends his specific research findings. By fostering an environment where theoretical research was applied practically, he cultivated a culture of scientific rigor within Austrian aviation research. This was particularly significant given Austria-Hungary's relative isolation from the main centers of aerodynamic research in Europe at that time.

Through his efforts, Vienna emerged as a foundational site of aerodynamic research, establishing a legacy that influenced future aeronautical advancements in Austria-Hungary and beyond. His contributions to aerodynamics before 1914 underscore a critical period of transition in aviation technology, advancing both the understanding of fluid dynamics and aeroplane design.[3]

Above: Richard Martin Edler von Mises, an Austrian mathematician and scientist, made groundbreaking contributions to aerodynamics and fluid mechanics. His work, integrating mathematical rigor with practical aviation challenges, played a crucial role in advancing flight theory and aeroplane design during the early 20th century.

Richard von Mises: Integrating Theory and Practice in Aerodynamic

Richard Martin Edler von Mises, born on 19 April 1883, in Lemberg, Austria-Hungary (now Lviv, Ukraine), was an Austrian scientist and mathematician renowned for his diverse contributions across several fields including solid mechanics, fluid mechanics, aerodynamics, aeronautics, statistics, and probability theory. His career in academia led him to the prestigious position of Gordon McKay Professor of Aerodynamics and Applied Mathematics at Harvard University.

Von Mises's contributions to aerodynamics were both practical and theoretical. He made significant advances in boundary-layer flow theory and airfoil design, crucial for understanding lift, drag, and other aerodynamic forces. His studies helped refine airfoil shapes for better aerodynamic efficiency and flight stability. Further enriching the scientific landscape, von Mises, a polymath whose expertise spanned mathematics, mechanics, and aerodynamics, developed theoretical frameworks and statistical methodologies that enhanced the precision of aerodynamic modeling and simulation. His rigorous approach to the fluid dynamics of air foreshadowed the complex computational techniques that would later become indispensable in aeronautical engineering.

Additionally, von Mises developed the distortion energy theory of stress, crucial in material strength calculations for understanding how materials deform under various types of stress, essential in designing aeroplane structures that are both light and durable.

Throughout his career, von Mises was known for his ability to apply rigorous mathematical principles to practical problems in engineering and physics. Beyond his technical achievements, von Mises was deeply engaged with the philosophy of science, aligning with neo-positivism and empiricism, and maintained intellectual exchanges with figures like Philipp Frank, contributing to the intellectual milieu that shaped early 20th-century scientific thought.

Richard von Mises's academic journey was marked by his early education in Vienna, where he excelled in Latin and mathematics, and his subsequent studies at the Vienna Polytechnic Institute. His doctoral work on "the determination of flywheel masses in crank drives" and his habilitation on "Theory of the Waterwheels" reflected his deep engagement with both theoretical and applied aspects of mechanics and engineering.

Before World War I, von Mises had already taken up flying, which naturally extended his academic interest into the realm of aeronautics. By 1913, at the University of Strasbourg, he delivered one of the first university courses on powered flight, highlighting his pioneering role in aeronautical education. His wartime service as a test pilot and flying instructor further exemplified his hands-on approach to aviation and his commitment to advancing flight technology.

Richard von Mises's contributions to aerodynamics and his broader scientific work reflect a profound capacity to integrate complex mathematical concepts with practical engineering challenges. His legacy in aerodynamics is marked by significant advancements in understanding airflow, stress in materials, and the optimal design of aeroplane structures, ensuring his lasting impact on the field.[4]

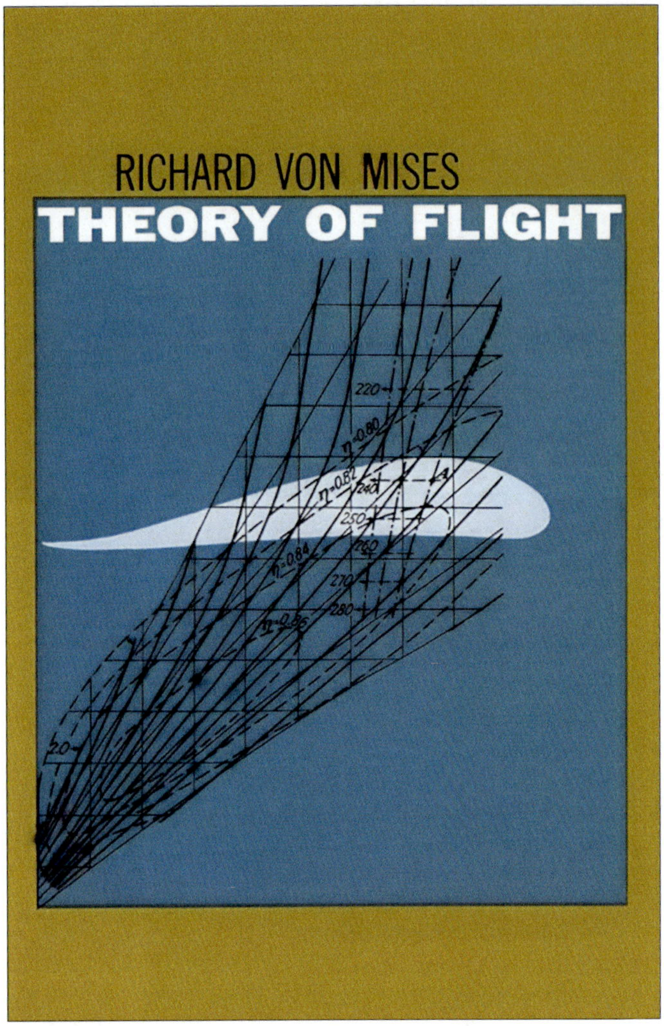

Above: Theory of Flight by Richard von Mises remains a seminal text in aeronautical engineering. This comprehensive study of aerodynamics, grounded in advanced mathematics, has left a lasting legacy on aviation technology and continues to influence the field of aerodynamics education.

Ernst Mach: Innovator in Aerodynamics

Ernst Mach, a pivotal figure in the scientific community of the late 19th and early 20th centuries, made groundbreaking contributions specifically in the field of aerodynamics that continue to influence modern aeronautical engineering. Best known for his exploration of shock waves and the introduction of the Mach number, Mach's work provided critical insights into supersonic motion, greatly enhancing the understanding and modeling of high-speed aerodynamics.

Born in Moravia in 1838, Mach's scientific journey led him through studies of physics at the University of Vienna, where his work initially focused on optics and acoustics. However, his interests and contributions extended significantly into the dynamics of gases and the effects of high speeds on objects moving through air. His profound investigations into the behaviors of objects traveling at speeds surpassing the speed of sound culminated in the development of the concept known as the Mach number. This dimensionless figure represents the ratio of an object's speed to the speed of sound in the surrounding medium and has become a fundamental concept in the study of aerodynamics and aeronautics.

Above: Ernst Mach, renowned physicist and innovator in the study of shock waves, whose work led to the development of the Mach number, a fundamental concept in aerodynamics for measuring supersonic speeds.

Mach's most notable aerodynamic experiments involved the use of schlieren photography to visualize shock waves generated by projectiles traveling at supersonic speeds. His meticulous studies not only confirmed the existence of shock waves but also described their conical shape, with the projectile at the apex. These observations were pivotal in understanding and designing more effective airfoils and improving the overall aerodynamic efficiency of aeroplanes.

Moreover, Mach's philosophical approach to science,

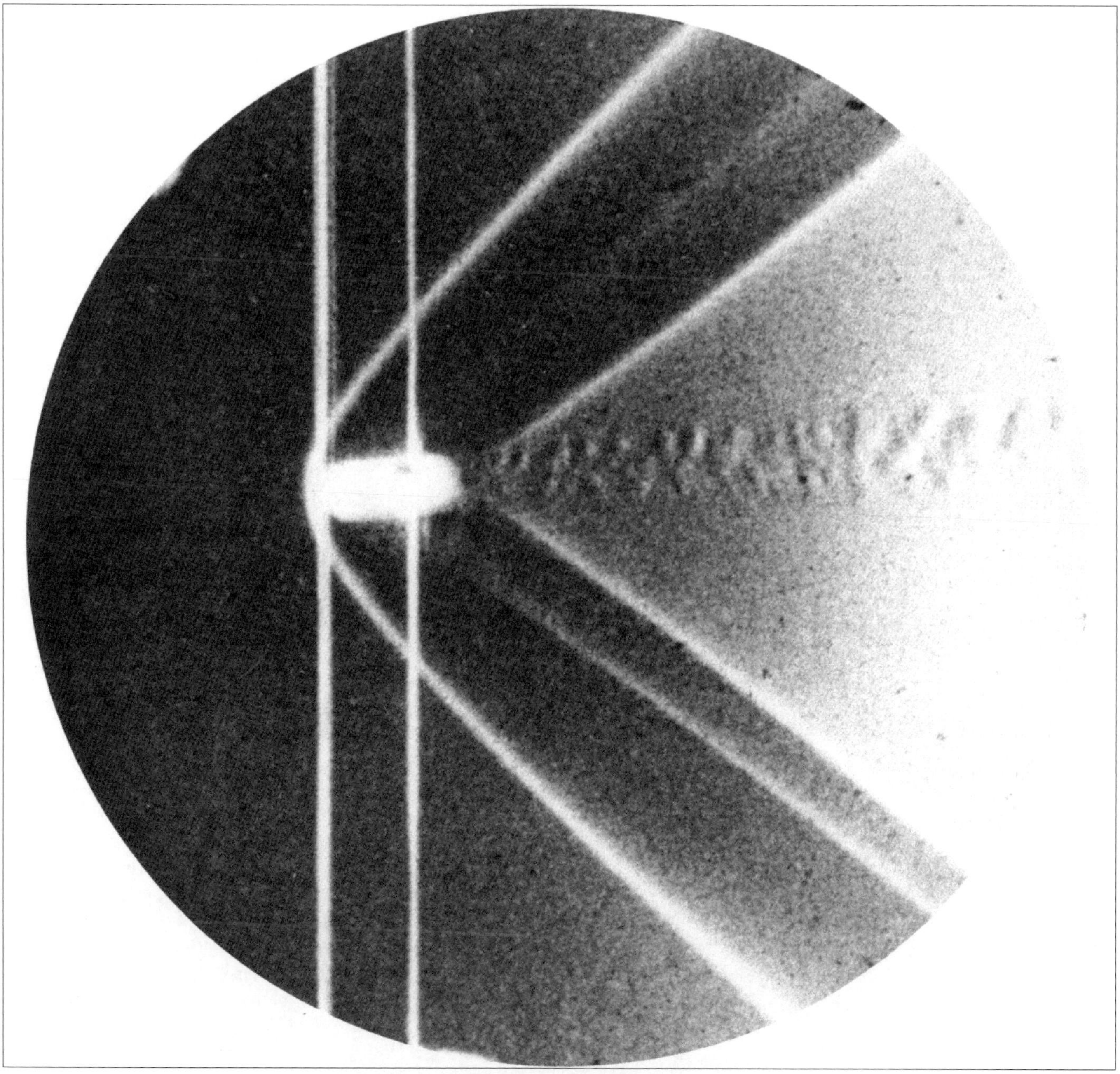

Above: A groundbreaking schlieren photograph taken by Ernst Mach in 1888, capturing the bow shock waves formed around a brass bullet traveling at supersonic speeds. Ernst Mach employed the schlieren technique, which utilizes the bending of light rays caused by variations in air density, to make invisible phenomena such as shock waves visible. By directing a focused light source through the region of interest and recording the refracted light patterns, Mach provided one of the earliest and most striking visualizations of high-speed aerodynamic flow. This pioneering work laid the foundation for understanding the behavior of objects at supersonic speeds and remains a cornerstone of fluid dynamics research.

particularly his critique of Newtonian mechanics and his discussions around the relative motion of objects, prefigured aspects of the theory of relativity. While his philosophical work extended beyond the direct boundaries of aerodynamics, his empirical methods and critical thought processes significantly influenced the way aerodynamic phenomena were conceptually understood and practically approached.

Ernst Mach's contributions to aerodynamics through his detailed study of supersonic motion and the establishment of the Mach number provided a fundamental understanding that has aided countless innovations in aeroplane design and performance. His legacy in aerodynamics is characterized by his rigorous scientific method and innovative approach to complex physical phenomena.[5]

Conspectus: The Emergence of Aerodynamics in Pre-WWI Austria-Hungary

In the vibrant scientific milieu of pre-WWI Austria-Hungary, a confluence of minds and institutions cultivated a pioneering spirit in the field of aerodynamics that resonated across the globe. This era, marked by rapid industrial and educational growth, fostered a unique synergy between theoretical insights and practical applications, profoundly shaping the development of modern aeronautics.

The significant contributions of Richard Knoller, Richard von Mises, and Ernst Mach exemplify the dynamic interplay of individual genius and collaborative academic environments that characterized the region's approach to scientific discovery. Knoller's work at the Vienna Polytechnic Institute, notably his development of the wind tunnel and the foundational theories of wingtip design, highlighted the practical integration of aerodynamic research with engineering education. His innovations not only advanced the understanding of fluid dynamics but also directly impacted aeroplane design and performance, enhancing the safety and efficiency of air travel.

Richard von Mises further enriched this landscape with his rigorous mathematical approaches to aerodynamics, particularly through his work on boundary-layer flow theory and the structural integrity of aeroplanes. His career bridged the gap between abstract mathematical concepts and their tangible applications in aeronautics, underscoring the importance of theoretical frameworks in solving complex engineering problems.

Ernst Mach, renowned for his philosophical and experimental contributions, brought a profound depth to the understanding of high-speed motion through his studies of shock waves and the Mach number. His work laid the groundwork for the scientific exploration of supersonic and hypersonic flight, which has become a cornerstone of contemporary aerospace research and design.

Together, these scholars contributed to a robust academic and industrial environment that propelled Austria-Hungary to the forefront of aerodynamic research during a critical period of technological transformation. Their legacy is evident in the continued relevance of their theories and methodologies in today's complex aeronautical engineering challenges.

The synergy between Austria-Hungary's educational institutions and its industrial ambitions created an ecosystem that was ideally suited for the flourishing of aerodynamics. The empire's technical schools and universities became hubs of innovation, where theoretical physics and practical mechanics converged to produce breakthroughs that extended the boundaries of what was possible in aviation. This period of intense scientific activity not only advanced the field of aerodynamics but also demonstrated the critical role of interdisciplinary collaboration in fostering technological innovation.

Endnotes

1. Kurt von Lössl: Der Lebenslauf Friedrich Ritter von Lössls [Kurt von Lössl: the curriculum vitae of Friedrich *Ritter* von Lössl]. In: *Schüt⊠enhofer V. (Hrsg.): Blätter für Technikgeschichte– Technikgeschichte in Wien,* vol 10. Springer, Wien, 1948.
2. Wilhelm Kress (1836–1913), an innovative aeronautical engineer and lecturer, is featured in a collection at the National Air and Space Museum Archives (Collection ID: NASM.XXXX.0862). Spanning the years 1880–1883, the collection showcases one of his German patents, Patent ⊠. 8706, for an early flying machine model. It also includes a letter from 1883 penned by Kress himself and an English description of the flying machine written by Arnold Samuelson, titled "Chief Engineer." Kress, born in Russia, ventured into aeronautics by initially inventing a motor-powered propeller-driven kite in 1865, known as the *Drachenflieger*. His career evolved through various innovations, including the design of a full-sized tandem-winged monoplane in 1897, culminating in his active participation in aeronautical lectures and writings until his demise in 1913. This repository is organized by type of material and offers insights into both the technical and historical facets of Kress's contributions to early aviation.
3. Hans Christoph Graf von Seherr-Thoß, "Knoller, Richard," *Neue Deutsche Biographie (NDB)* [The new German biographical reference], vol. 12, (Berlin: Duncker & Humblot, 1980), 210–211. ISBN 3-428-00193-1; Richard Knoller, *Vorlesungen über Luftschiffahrt* [Airship lectures], 2nd ed. (Vienna: Stiw, 1920), 89 pp. with VII plates. 1921 B 677, Leipzig: Richard Knoller, *Die Gesetze des Luftwiderstandes* [The laws of air resistance], Verlag des Österreichischer Flugtechnischen Vereines, 1909: Richard Knoller, *Flüssigkeitswiderstand und Propellertheorie* [Liquid resistance and propeller theory], Verlag des Österreichischen Flugtechnischen Vereines, 1909.
4. Richard von Mises, *Fluglehre: Theorie u. Berechnung d. Flugzeuge in elementarer Darstellung* [The flight instructor: fundamental theory and assessment of the aeroplane], Berlin, Springer, 1957.
5. Rudolf Dvořák, "Ernst Mach in Prague and the Dawn of Gas Dynamics," in *Ernst Mach–Life, Work, Influence*, ed. Friedrich Stadler, Springer, 2019, 128–129.

11. The Austro-Hungarian Flight Journey

Left: The Austro-Hungarian Monarchy's deep connection with aviation is symbolized in this image, with Archduke Josef Ferdinand (left) and Archduke Heinrich Ferdinand (right) standing beside Franz Hinterstoisser in the balloon gondola. All three were enthusiastic aeronauts, embarking on numerous balloon flights as a symbol of their dedication to aviation exploration. This moment reflects the Monarchy's embrace of aviation, driven by a commitment to technological innovation and military preparedness. The Austro-Hungarian fascination with ballooning laid the groundwork for the emergence of its military aviation capabilities, evolving from balloon flights to the more strategic use of aeroplanes as the Empire recognized the potential of the skies for both exploration and warfare. This image encapsulates the continuity of Austria-Hungary's commitment to aviation, from pioneering balloon flights to the establishment of its military air power.

Austria-Hungary's military forces arrived relatively late to the domain of aviation, with the initial pioneering activities primarily driven by individuals from the civilian sector. However, the lines between civilian and military spheres within Austro-Hungarian society were frequently blurred. Some civilian and military pilots pursued flight training abroad, returning to their homeland to disseminate the skills and insights they had acquired. While these cross-border educational experiences were beneficial, the nature of flight training was predominantly indigenous, thus profoundly shaping the early aviation landscape in Austria-Hungary. These aviators participated in air meets, set records, and won prizes, reflecting the period's intense national competition in aviation, with various records and accolades at stake across numerous contests. By the end of 1912, Austro-Hungarian pilots, employing both domestically produced and foreign aeroplanes, had secured a number

Above: *FliK* 7 personnel with an Aviatik B.II at Nisko am San, August 1914. From left to right: *Oblt* Artur Kollitsch, *Oblt* Karl Banfield, *Oblt* Franz Hellmann, *Hptm* Franz Smicka, *Hptm* Robert von Kaiserfeld, *Hptm* Franz Schorn, *Hptm* Richard Schuster, *Oblt* Maximillian Perini, and *Hptm* Josef Bendik. The group comprises both pilots and observers, reflecting the varied responsibilities within the unit. *Credit: Christian Reiterer*

of world records.[1]

Viktor Silberer's role as a pioneer and advocate for aeronautics significantly shaped public and military perspectives on aviation. His efforts culminated in the integration of balloon-based services into the *k.u.k. Armée*, setting the stage for further aeronautical advancements. At Fischamend, technological innovations such as the early use of aerial photography and the transition toward aeroplane-based military services highlighted the strategic shift in Austro-Hungarian aviation strategy. This period also marked a crucial point in military aviation, influenced heavily by Franz Conrad von Hötzendorf, whose advocacy for the use of aeroplanes in military strategy was pivotal. His efforts ensured that Austria-Hungary, though initially lagging behind its neighbors, began to recognize and harness the potential of advanced aeronautical technology for military applications.

The incorporation of balloon-based aeronautic services into the *k.u.k. Armée* occurred in 1893, seven years after Germany, while the decision to establish aeroplane-based services was wholly undertaken in 1912, two years after Germany instituted its provisional military flying training facility at Döberitz in 1910. The complex decision-making processes of the Habsburg multi-ethnic state administration further complicated matters, as military expenditures required laborious political compromises between the Austrian and Hungarian segments of the empire.[2]

Despite the *k.u.k.* Monarchy's belated development of aviation compared to other major European powers, both *k.u.k.* military pilots and Austro-Hungarian civilian pilots achieved noteworthy successes in aviation competitions and flight records. As World War I commenced in August 1914, military aviation had progressed more slowly compared to other prominent

Above: *Linienschiffsleutnants* [Lieutenants] Hugo Ockermüller (pilot, left) and *Beobachter* [Observer] Heinrich Fontaine von Felsenbrunn prepare for takeoff on 2 October 1914 from the *Seeflugstation* in Kumbor, located in the strategically significant Bay of Kotor. Their mission—a bombing raid targeting Budua (modern-day Budva, Montenegro)—represented one of the earliest operational uses of Austro-Hungarian naval aviation in World War I. Flying the Lohner flying boat "Type M," designated E18, they demonstrated the growing coordination between air and naval forces in a complex maritime theater. With Montenegro as a critical adversary in the Adriatic, such missions highlighted the emerging role of aerial reconnaissance and tactical bombing in Austro-Hungarian naval strategy.

European powers. The Imperial war ministers perpetually faced financial constraints more severe than their French, Russian, or Prussian-German counterparts, despite the Danube Monarchy's vast size and manpower potential. This disadvantage was especially pronounced for the airmen but was also experienced by military planners and high-ranking officers across all armies.

It is essential to recognize that most war historians evaluate these decisions with the advantage of hindsight. In their time, it was uncertain whether the allocated financial resources and military equipment would guarantee success. For example, Germany's investment in the zeppelin airship fleet illustrates how armament funds were allocated to expensive and ineffective weapon systems, despite seemingly reasonable operational reasons to the contrary. In the airship sector, the Vienna War Ministry avoided such disappointment. However, it remains clear that financial constraints were the deciding factor for the *k.u.k.* War Ministry's decision against acquiring airships, despite the domestic industry's willingness and investment readiness to support their construction. It seems more rational to view that the military was more invested in developing an aeroplane-based service, rather than allocating the limited available funds for zeppelin-type airships. This choice reflects a strategic preference for the more versatile and potentially cost-effective aeroplane technology over airships.[3]

On 30 June 1914, the *k.u.k.* War Ministry decided

to establish 15 *FliK*s with a total of 270 aeroplanes. To achieve this number, 109 aeroplanes were to be purchased, and ground support facilities in Fischamend, Wiener Neustadt, and elsewhere were to be expanded. Although fully operational airfields had not yet been established at Goerz, Krakau, and Przemyśl, preparations were in progress. Orders were placed for 24 ready-to-deliver aeroplanes, with the remainder scheduled for purchase by November or December. Nevertheless, the war's onset disrupted these initial plans. By the end of 1914, 173 aeroplanes were received in total, including 91 domestically produced and 82 from Germany. This inventory allowed for the establishment of nine *FliK*s in October, one in September, and four on 2 November, with 14 ultimately deployed. Simultaneously, extensive testing with photographic and radio equipment, as well as weaponry, took place. The continuous limitation of domestic production and the reliance on German imports for aeroplanes, engines, and related equipment would persistently hinder the *k.u.k. Luftschifferabteilung's* [Aeronautics Department's] ability to fulfill numerous tasks necessary to support the war effort. Nonetheless, the effort and dedication of those involved made a remarkable difference in the overall performance of the Austro-Hungarian aviation forces.[4]

Despite the setbacks and challenges, the *k.u.k.* Aeronautics Department's personnel managed to adapt and evolve, maximizing the use of available resources. They developed innovative tactics and strategies that contributed to the Austro-Hungarian military's overall performance during the war. As the conflict progressed, the *k.u.k.* Aeronautics Department continued to refine its aerial reconnaissance, artillery spotting, and aerial combat capabilities. They also engaged in experimental projects, including the development of early bomber aeroplanes and the implementation of wireless telegraphy and aerial photography for intelligence purposes.[5]

In conclusion, although the Austro-Hungarian aviation forces may have entered the field later than their European counterparts, they exhibited significant advancements and successes in both civilian and military sectors. The complex political landscape and financial constraints of the Austro-Hungarian Empire hampered the development and growth of its aviation capabilities. Nevertheless, the commitment and ingenuity of its pilots, engineers, and support staff contributed to overcoming some of these limitations. As a result, the *k.u.k.* Aeronautics Department and, later, the *k.u.k. Luftfahrtruppen* [Air Troops] played a crucial role in the Austro-Hungarian military's performance during World War I, despite the numerous obstacles they faced.

Endnotes

1. In 1912, the landscape of aviation was marked by intense competition among pilots worldwide, striving to set and break records across various categories. French pilots led the tally with 64 world records, primarily in speed, benefiting from their light and agile aeroplane designs. Austrian pilots, utilizing the robust Austro-Daimler engines, excelled in endurance and altitude, securing 20 world records. In contrast, German pilots claimed four records, while Belgian and British pilots each achieved one. The distribution of pilots also reflected national priorities in aviation: France boasted 614 pilots, representing 48% of the world's total pilot population of 1,285, underscoring its leadership in aviation development. Austria, with 42 pilots (3%), Germany with 135 (11%), Belgium with 48 (4%), and Great Britain with 200 (16%), demonstrated varying levels of investment and focus on aviation capabilities.
2. Zelezny, *Neustadt*, 147–148.
3. Desoye, 52, 60, 227.
4. Desoye, 177–179.
5. Steinböck, *Bewaffnung*, 7–59.

BIBLIOGRAPHY

Primary Sources:

Albrich, Thomas, and Nikolaus Hagen (eds.). *Austria-Hungary's Flying Aces in the First World War 1914–1918.* Universitätsverlag Wagner, Innsbruck, 2019.

BMLVS (ed.). *The Third Dimension. Military Aviation Exhibition Zeltweg.* Vienna, 2013.

Chant, Christopher. *Osprey Aircraft of the Aces - Austro-Hungarian Aces of World War 1.* Osprey Publishing, Oxford, 2002.

Conrad von Hötzendorf. *Aus meiner Dienstzeit 1906-1918 Band II* [From My Service 1906-1918 Volume II], Rikola Verlag, Wien, 1922.

Červenka, Juraj. *Habsburští orli 1. Rakousko-uherské letectvo, jeho esa a protivníci v letech 1914-1918.* Český cestovatel.cz, Praha, 2015.

Červenka, Juraj. *Habsburští orli 2. Rakousko-uherské letectvo, jeho esa a protivníci v letech 1914-1918.* Český cestovatel.cz, Praha, 2015.

Davilla, James J., and Arthur M. Soltan. *French Aircraft of the First World War.* Flying Machines Press, Stratford, 1997.

Deák, István. *Beyond Nationalism. A Social and Political History of the Habsburg Officers Corps, 1848-1918.* Oxford University Press, New York – Oxford, 1992.

Desoye, Reinhard Karl Boromäus. *Die k.u.k. Luftfahrtruppe - Die Entstehung der Aufbau und die Organisation der österreichisch-ungarischen Heeresluftwaffe 1912-1918.* Diplomica Verlag, Hamburg, 2000.

Düsing, Michael. *German & Austro-Hungarian Aero Engines of WWI, Vol. 1-3.* Aeronaut Books, 2022.

Fink, Carl. *Die Entwicklung des militärischen deutschen Luftbildwesens 1911-1918 und seine militärische wie kulturelle Bedeutung.* Wehrwissenschaftliche Rundschau, Jg. 10, Heft 7 (1960) 390-399.

Fink, Carl. *Einiges von der Entwicklung unserer Flugzeugkameras.* Illustrierte Flug-Welt, (Juni 1920).

Fink, Carl. *Das Reihenbild als Kartenersatz.* Illustrierte Flug-Welt, (Mai 1920).

Grosz, Peter, George Haddow, and Peter Schiemer. *Austro-Hungarian Army Aircraft of World War One.* Flying Machines Press, Mountain View, 1993.

Hauke, Erwin, Walter Schroeder, and Bernhard Tötschinger. *The Aircraft of the kuk Aviation Troops and Naval Aviation 1914–1918.* H. Weishaupt Verlag, Graz, 1988.

Hoernes, Hermann. *Buch des Fluges. III. Band* [Book of Flight. Vol. III], Wien 1912. Verlag Georg Szelinski, K.K. Universitäts-Buchhandlung.

Jäger, Helmut. *Erkundung mit der Kamera: die Entwicklung der Photographie zur Waffe und ihr Einsatz im 1. Weltkrieg.* München, Venorion VKA, 2007.

Keimel, Reinhard. *Flugzeuge der österreichischen Firma Lohner, 1909-1923.* Sonderheft der Blätter für Technikgeschichte, 1990.

Keimel, Reinhard. *Österreichs Luftfahrzeuge: Geschichte der Luftfahrt von den Anfängen bis Ende 1918.* H. Weishaupt Verlag, 1981.

Kerchnawe, Hugo. *Die unzureichende Kriegsrüstung der Mittelmächte als Hauptursache ihrer Niederlage,* Ergänzungsheft 4 zum Werke 'Österreich-Ungarns Letzter Krieg', Militärwissenschaftlichen Mitteilungen, Wien, 1932.

Knoller, Richard. *Die Gesetze des Luftwiderstandes.* Verlag des Österreichischer Flugtechnischen Vereines, 1909.

Knoller, Richard. *Flüssigkeitswiderstand und Propellertheorie.* Verlag des Österreichischen Flugtechnischen Vereines, 1909.

Knoller, Richard. *Vorlesungen über Luftschiffahrt.* 2nd ed. (Vienna: Stiw, 1920).

Meindl, Karl. *Luftsiege der k.u.k. Luftfahrtruppen.* Volumes 1-8, ÖFH-Nachrichten - Sonderheft, Wien, 2001-2007.

Morrow, John Howard. *The Great War in the Air: Military Aviation From 1909 to 1921.* University of Alabama Press, Tuscaloosa, Ala., 2009.

Neumann, Georg Paul. *Die deutschen Luftstreitkräfte im Weltkriege.* Berlin: E. S. Mittler und Sohn, 1920.

O'Connor, Martin. *Air Aces of the Austro-Hungarian Empire 1914-1918.* Champlin Fighter Museum Press, 1986.

Olejko, Andrzej. *The Habsburgs' Wings 1914.* Library of Armed Conflicts, Kagero, 2018.

Pandis, Robert S. *Flight Badges of the Central Powers 1914-1918, Volume II: The Imperial German Naval Austro-Hungarian Bulgarian and Ottoman Turkish Air Services.* Imperial House Antiques LLC, 2021.

Peter, Ernst. *Die k.u.k. Luftschiffer und Fliegertruppe Österreich-Ungarns 1794-1919.* Motorbuch, Stuttgart, 1981.

Plattner, Peter, et al. *Die k.u.k. aeronautischen Anstalt Fischamend. Der aviatische Salon.* Wien, 2011.

Ravbar, Matjaž. *Avstro-Ogrsko letalstvo na soški fronti 1915–1917.* Cesarska in kraljeva letališka infrastructurea v zaledju soške fronte. Univerza, Ljubljana, 2011.

Richter, Oliver. *Feldluftschiffer: The German Balloon Corps and Aerial Reconnaissance.* Erlangen Germany: Verlag Jochen Vollert Tankograd Publishing, 2013.

Rauchensteiner, Manfried. *The First World War and the End of the Habsburg Monarchy 1914-1918.* Böhlau, Vienna, 2014.

Ringl, Reinhard, and Rudolf Ster. *Die k.u.k. Aeronautische Anstalt Fischamend. Band 1. Die grosse Zeit der K.u.K. Luftschiffe 1908-1914.* Carinaverlag, 2017.

Schupita, Peter. *Die k.u.k. Seeflieger: Chronik und Dokumentation der österreichisch-ungarischen Marineluftwaffe 1911-1918.* Bernard & Graefe Verlag, Koblenz, 1983.

Silberer, Victor. *Im Ballon! Eine Schilderung der Fahrten des Wiener Luftballons 'Vindobona' im Jahre 1882.* Verlag der Allgemeinen Sport-Zeitung, Wien, 1883.

Sore, Piero. *Storia dell'Aviazione di Marina Austro-Ungarica*

1909–1918. Aviani & Aviani editori, 2023.

Stahl, Friedrich-Christian. *Die Militärluftfahrt bis zum Beginn des Weltkriegs 1914.* Mittler, Berlin, 1966 (Reprint from Orig. 1941).

Steinböck, Erwin. *Bewaffnung und Ausrüstung der österreichisch-ungarischen Flugzeuge (Österreichs Luftfahrt in Einzeldarstellungen).* H. Weishaupt, 1983.

Steiner, Jörg C. *Das Feldpiloten-Abzeichen, Militärhistorische Themenreihe Bd. 4.* Wien: Edition S & H, 1992.

Vego, Milan N. *Austro-Hungarian Naval Policy 1904–14.* Frank Cass, London, 1996.

Waffen-Arsenal Band-149 Deutsche Fessel und Sperrballone 1900-1945 [German Tethered and Barrage Balloons 1900-1945].

Zelezny, Marcus. *Aviatik in Wiener Neustadt (1909-1912). Geschichte Auf Postkarten: Aviation in Wiener Neustadt (1909-1912). History on Postcards.* [bilingual German/English]. Stanger Verlag, 2022.

Zelezny, Marcus. *Pilotendiplome des Österreichischen Aero-Clubs. Inhaber, Prüfer und Bedingungen (1910-1919), Pilot diplomas of the Austrian Aero-Club – Owners, commissioners & regulations (1910-1919)* [bilingual German/English]. Vienna, 2019.

Zelezny, Marcus & Mayrhofer, Koloman. *Borduhren – Cockpitausrüstung österreichisch-ungarischer Flugzeuge, Cockpit clocks – Cockpit equipment of Austro-Hungarian aeroplanes* [bilingual German/English]. Vienna, 2019.

Zelezny, Marcus & Reiterer, Christian. *Patriotische Fliegerabzeichen. Abzeichen und Memorabilien der österreichischen Luftfahrt (1882-1918), Patriotic aviation badges – Badges and Memorabilia of Austrian Aviation (1882-1918)* [bilingual German/English]. Vienna, 2019.

Periodicals, Journals and Newspapers:

Cross & Cockade Journal (US). *1960–1986* (Cross & Cockade Journal, United States) Dedicated to the history of World War I aviation, this journal offered scholarly articles, personal accounts, and research findings pertinent to early aviation history.

Cross & Cockade International (originally Cross & Cockade Great Britain). *United Kingdom, since 1970* (Journal dedicated to the study and preservation of World War I aviation history) Published by Cross & Cockade International, it features in-depth articles, technical details, and historical analysis specific to early military aviation.

Deutsche Luftfahrer-Zeitung. Berlin, 1912–1919 (German Aviators' Newspaper, Berlin, 1912–1919). A leading publication of its time, it provided comprehensive coverage of aviation news, technological advancements, and pilot experiences within Germany and internationally.

Flug- und Motortechnik, Jg. 1909/10. *Offizielle Mitteilungen des Österreichischen Flugtechnischen Vereines,* Wien, seit 1908 (Flight and Motor Technology, Annual 1909/10. Official Notices of the Austrian Aerotechnical Association, Vienna, since 1908). This publication served as the official communication organ of the Austrian Aerotechnical Association, offering technical articles, research findings, and updates on advancements in flight and motor technology during the early 20th century.

Fluginformationen des Österreichischen Luftfahrt-Archives. Wien, seit 1950 (Flight Information of the Austrian Aviation Archives, Vienna, since 1950). This publication provided updates, research articles, and historical accounts related to Austrian aviation, serving as a resource for historians and enthusiasts.

Flugsport – Illustrierte technische Zeitschrift und Anzeiger für das gesamte Flugwesen. Frankfurt a. M., seit 1908 (Aviation – Illustrated Technical Journal and Announcer for the Entire Aviation Field, Frankfurt a. M., since 1908). Founded by Oskar Ursinus, this bi-weekly magazine was instrumental in disseminating technical knowledge and fostering a community among aviation enthusiasts and professionals.

Illustrierte Aeronautische Mitteilungen. Straßburg/Berlin, 1897–1909 (Illustrated Aeronautical Reports, Strasbourg/Berlin, 1897–1909). This publication featured illustrated reports on aeronautics, including ballooning and early powered flight, contributing to the popularization and technical understanding of aviation.

Jahresberichte des Wiener Aero-Klub. Wien, 1901–1910 (Annual Reports of the Vienna Aero Club, Vienna, 1901–1910). These annual reports documented the activities, events, and developments within the Vienna Aero Club, offering insights into the early aviation community in Austria.

Jahrbücher des k.k. Österreichischen Aero-Clubs. Wien, ab 1911 (Yearbooks of the Imperial and Royal Austrian Aero Club, Vienna, from 1911). Serving as comprehensive records, these yearbooks detailed the progress, events, and contributions of the Austrian Aero Club to the field of aviation.

Kölnische Zeitung. Cologne, 1802–1945 (Cologne Newspaper). One of Germany's oldest newspapers, it included coverage of aviation events and developments, reflecting the public's growing interest in flight.

OFH-Nachrichten, Informationsblatt der Österreichischen Flugzeughistoriker. Wien, seit 1977 (OFH News, Information Bulletin of Austrian Aircraft Historians, Vienna, since 1977). As the information bulletin of Austrian aircraft historians, this periodical offered articles, research findings, and discussions on the history of aircraft and aviation in Austria.

Österreichische Automobil-Zeitung. Wien, 1900–1938 (Austrian Automobile Newspaper, Vienna, 1900–1938). Initially focused on automotive topics, this publication expanded to cover aviation advancements, reflecting the growing intersection of automobile and flight technologies in the early 20th century.

Österreichische Flug-Zeitschrift. *Hrsg. vom k.k. Österreichischen Flugtechnischen Verein,* Wien, seit 1911 (Austrian Flight Journal, published by the Imperial and Royal Austrian Aerotechnical Association, Vienna, since 1911). As the official journal of the Imperial and Royal Austrian Aerotechnical Association, this periodical provided comprehensive coverage of aviation developments, including technological innovations, pilot experiences, and industry news within Austria and beyond.

Österreichischer Motor – Der Flug. Wien, seit 1914 (Austrian Motor – The Flight, Vienna, since 1914). This periodical provided insights into both automotive and aviation advancements, highlighting the technological progress and integration of motorized vehicles and aircraft in Austria.

Over the Front. *United States, since 1986* (Journal of the League of WWI Aviation Historians) This publication offers in-depth studies, technical articles, and first-hand accounts from World War I, exploring various aspects of early military aviation and its legacy.

Wiener Luftschiffer-Zeitung. Wien, 1902–1914 (Vienna Aeronaut Newspaper, Vienna, 1902–1914). Focused on lighter-than-air flight, this publication covered developments in ballooning and airship technology, documenting the evolution of aerial navigation before the widespread adoption of airplanes.

Zeitschrift des Vereines zur Förderung der Luftschiffahrt. Berlin, 1882–1903 (Journal of the Association for the Promotion of Airship Navigation, Berlin, 1882–1903). Dedicated to promoting airship navigation, this journal featured articles on airship design, engineering challenges, and the potential applications of lighter-than-air craft during the late 19th and early 20th centuries.

Archival and Manuscript Sources:

Österreichisches Bundesministerium für Heereswesen / Kriegsarchiv, Wien, Austria – Repository of Austrian military records, particularly on WWI aviation strategies, air force operations, and technical innovations within Austro-Hungarian military aviation.

Militärwissenschaftliche Mitteilungen [Military Science Communications]. Vienna, 1931 – A publication by the Austrian Ministry of Defense covering advancements in military science and technology, relevant to Austrian and Austro-Hungarian forces.

Österreich-Ungarns Letzter Krieg, 1914-1918. 2nd ed. Vienna, 1931 – A multi-volume series detailing Austria-Hungary's military engagements in WWI. This includes:

Erster Band [First Volume] – Overview of early WWI campaigns.

Erster Band, Beilagen [First Volume, Annexes] – Additional documents and illustrations for Volume 1.

Zweiter Band [Second Volume] – Further account of Austria-Hungary's military operations.

Zweiter Band, Beilagen [Second Volume, Annexes] – Supplementary materials for Volume 2.

Glaise-Horstenau, Edmund, ed. Österreich-Ungarns Letzter Krieg, 1914-1918, Volumes 1 & 2. Translated by Stan Hanna. Vienna: Publisher of Military Science Releases, 1930-1931 – Edited volumes with an English translation, detailing significant campaigns in 1914 and 1915.

Fernbrugg, Fiala von. *Radio-Artill. Schiessen* [Radio Artillery Shooting]. MS/Lu Nr. 12 – Manuscript on early radio-guided artillery tactics and technology in Austro-Hungarian military applications.

Regele, Oskar. *Die Tätigkeit der österreichisch-ungarischen Luftstreitkräfte beim Auffangen der russischen Angriffsheere im Sommer und Herbst 1914,* in *Die österreichisch-ungarischen Luftstreitkräfte im Weltkriege 1914-18,* Volume 2. Ed. Kriegswissenschaftliche Abteilung der Luftwaffe, Zweigstelle Wien, 1942/43 – Documents the role of Austro-Hungarian air forces against Russian advances, with supplementary folders included.

Regele, Oskar. *Die Tätigkeit der österreichisch-ungarischen Luftstreitkräfte im Feldzuge gegen Serbien und Montenegro im Jahre 1914,* in *Die österreichisch-ungarischen Luftstreitkräfte im Weltkriege 1914-18,* Volume 1. Ed. Kriegswissenschaftliche Abteilung der Luftwaffe, Zweigstelle Wien, 1941/42 – Examination of air operations against Serbia and Montenegro, including annex volumes.

Peters, Klaus. *Zur Entwicklung der österreich-ungarischen Militärluftfahrt* [On the Development of Austro-Hungarian Military Aviation] – Graduate thesis exploring the evolution of Austro-Hungarian aviation forces. Archived in the Austrian State Archives under Document ID: 59.894.

Other Archives:

Austrian State Archives, Österr. Staatsarchiv II, Document ID: 59.894 – Contains official documents and records related to Austrian military aviation history, including research material on the development of Austro-Hungarian military aviation.

Kriegsarchiv Wien, Das Lichtbildwesen der öst.-ungarischen Luftfahrtruppen 1914/18 – Typographic manuscript from the Austrian War Archives focusing on the photographic documentation practices of Austro-Hungarian air troops during WWI.

National Air and Space Museum Archives, Wilhelm Kress Collection, NASM.XXXX.0862, 1880-1883 – This archive focuses on Wilhelm Kress's pioneering aeronautical experiments and designs, providing insight into the technological foundations of early aviation. The museum's broader archival collection includes extensive materials on aviation history, with early aeronautical engineering documents, photographs, and technical drawings. This valuable resource supports in-depth research into the development of flight technology and significant historical advancements in aviation.

Ku.k. Kriegsmarine Archives, Observation Balloon Tests and Exercises 1902-1907 – Records documenting early Austro-Hungarian naval observation balloon activities, with technical details on balloon testing and exercises.

Interessengemeinschaft Luftfahrt Fischamend, Correspondence with Ing. Rudolf Ster – Collection of early aviation-related correspondence and materials from Luftfahrt Fischamend, 1909-1913, highlighting contributions from engineers such as Rudolf Ster.

Library of Congress Collection on Franz Hinterstoisser. *Aus meinem Luftschiffertagebuche.* Leipzig: Verlag von H. Haessel, 1904 – Rare collection of Hinterstoisser's diary, providing personal accounts of early airship navigation and military applications. Located in the Library of Congress, Rare Book and Special Collections Division, Washington, D.C.

INDEX

Individuals

Aeroplanes, Airships, Balloons, & Equipment

Companies & Associations

Military Commands and Government Offices

Events

Made in United States
Orlando, FL
04 November 2026

90206123R00088